Common Core Math Practice Grade 7

Complete Content Review Plus 2 Full-length Common Core Math Tests

Elise Baniam - Michael Smith

Common Core Math Practice Grade 7

Common Core Math Practice Grade 7
Published in the United State of America By
The Math Notion
Email: info@Mathnotion.com
Web: WWW.MathNotion.com

Copyright © 2020 by the Math Notion. All rights reserved. No part of this publication may be reproduced, stored in a retrieval system, or transmitted in any form or by any means, electronic, mechanical, photocopying, recording, scanning, or otherwise, except as permitted under Section 107 or 108 of the 1976 United States Copyright Ac, without permission of the author.
All inquiries should be addressed to the Math Notion.

ISBN: 978-1-63620-035-4

About the Author

Elise Baniam has been a math instructor for over a decade now. She graduated in Mathematics. Since 2006, Elise has devoted his time to both teaching and developing exceptional math learning materials. As a Math instructor and test prep expert, Elise has worked with thousands of students. She has used the feedback of her students to develop a unique study program that can be used by students to drastically improve their math score fast and effectively.

– SAT Math Workbook
– ACT Math Workbook
– ISEE Math Workbooks
– SSAT Math Workbooks
–many Math Education Workbooks
– and some Mathematics books …

As an experienced Math teacher, Mrs. Baniam employs a variety of formats to help students achieve their goals: she teaches students in large groups, and she provides training materials and textbooks through her website and through Amazon.

You can contact Elise via email at:
Elise@Mathnotion.com

Common Core Math Practice Grade 7

Get the Targeted Practice You Need to Excel on the Math Section of the Common Core Test Grade 7!

Common Core Math Practice Grade 7 is **an excellent investment in your future** and the best solution for students who want to maximize their score and minimize study time. Practice is an essential part of preparing for a test and improving a test taker's chance of success. The best way to practice taking a test is by going through lots of Common Core math questions.

High-quality mathematics instruction ensures that students become problem solvers. We believe all students can develop deep conceptual understanding and procedural fluency in mathematics. In doing so, through this math workbook we help our students grapple with real problems, think mathematically, and create solutions.

Common Core Math Practice Book allows you to:

- Reinforce your strengths and improve your weaknesses
- Practice **2500+ realistic** Common Core math practice questions
- Exercise math problems in a variety of formats that provide intensive practice
- Review and study **Two Full-length Common Core Practice Tests** with detailed explanations

...and much more!

This Comprehensive Common Core Math Practice Book is carefully designed to provide only that **clear and concise information** you need.

WWW.MathNotion.com

… So Much More Online!

- ✓ FREE Math Lessons

- ✓ More Math Learning Books!

- ✓ Mathematics Worksheets

- ✓ Online Math Tutors

For a PDF Version of This Book

Please Visit WWW.MathNotion.com

Common Core Math Practice Grade 7

Contents

Chapter 1: Whole Numbers 11
    Add and Subtract Integers 12
    Multiplication and Division 13
    Absolute Value 14
    Ordering Integers and Numbers.. 15
    Order of Operations 16
    Factoring 17
    Great Common Factor (GCF) 18
    Least Common Multiple (LCM) 19
    Divisibility Rule 20
    Answer key Chapter 1.. 21

Chapter 2: Fractions 25
    Adding Fractions – Like Denominator 26
    Adding Fractions – Unlike Denominator 27
    Subtracting Fractions – Like Denominator 28
    Subtracting Fractions – Unlike Denominator 29
    Converting Mix Numbers 30
    Converting improper Fractions 31
    Addition Mix Numbers 32
    Subtracting Mix Numbers 33
    Simplify Fractions 34
    Multiplying Fractions 35
    Multiplying Mixed Number 36
    Dividing Fractions 37
    Dividing Mixed Number 38
    Comparing Fractions 39
    Answer key Chapter 2.. 40

Chapter 3: Decimal 45
    Round Decimals 46
    Decimals Addition 47
    Decimals Subtraction 48

Common Core Math Practice Grade 7

 Decimals Multiplication 49

 Decimal Division 50

 Comparing Decimals 51

 Convert Fraction to Decimal 52

 Convert Decimal to Percent 53

 Convert Fraction to Percent 54

 Answer key Chapter 3 55

Chapter 4: Exponent and Radicals 58

 Positive Exponents 59

 Negative Exponents 60

 Add and subtract Exponents 61

 Exponent multiplication 62

 Exponent division 63

 Scientific Notation 64

 Square Roots 65

 Simplify Square Roots 66

 Answer key Chapter 4 67

Chapter 5: Ratio, Proportion and Percent 70

 Proportions 71

 Reduce Ratio 72

 Percent 73

 Discount, Tax and Tip 74

 Percent of Change 75

 Simple Interest 76

 Answer key Chapter 5 77

Chapter 6: Measurement 79

 Reference Measurement 80

 Metric Length Measurement 81

 Customary Length Measurement 81

 Metric Capacity Measurement 82

 Customary Capacity Measurement 82

 Metric Weight and Mass Measurement 83

 Customary Weight and Mass Measurement 83

 Unit of Measurements 84

WWW.MathNotion.com

Temperature 85

Time 86

Answers of Worksheets – Chapter 6 87

Chapter 7: Linear Functions 89

Relation and Functions 90

Slope form 91

Slope and Y-Intercept 91

Slope and One Point 92

Slope of Two Points... 93

Equation of Parallel and Perpendicular lines... 94

Quadratic Equations - Quadratic Formula... 95

Answer key Chapter 7... 96

Chapter 8: Equations and Inequality 98

Distributive and Simplifying Expressions 99

Factoring Expressions 100

Evaluate One Variable Expressions 101

Evaluate Two Variable Expressions 102

Graphing Linear Equation 103

One Step Equations... 104

Two Steps Equations 105

Multi Steps Equations 106

Graphing Linear Inequalities 107

One Step Inequality 108

Two Steps Inequality 109

Multi Steps Inequality.. 110

Finding Distance of Two Points 111

Answer key Chapter 8... 112

Chapter 9: Transformations 117

Translations 118

Reflections 119

Rotations 121

Dilations 123

Coordinates of Vertices 124

Answers of Worksheets – Chapter 9 125

Chapter 10: Geometry 129
Area and Perimeter of Square 130
Area and Perimeter of Rectangle 131
Area and Perimeter of Triangle...132
Area and Perimeter of Trapezoid 133
Area and Perimeter of Parallelogram 134
Circumference and Area of Circle 135
Perimeter of Polygon 136
Volume of Cubes 137
Volume of Rectangle Prism 138
Volume of Cylinder 139
Volume of Spheres 140
Volume of Pyramid and Cone..141
Surface Area Cubes 142
Surface Area Rectangle Prism 143
Surface Area Cylinder 144
Answer key Chapter 10..145

Chapter 11: Statistics and probability 147
Mean, Median, Mode, and Range of the Given Data 148
Box and Whisker Plot 149
Bar Graph 150
Dot plots..151
Scatter Plots 152
Stem–And–Leaf Plot 153
Pie Graph 154
Probability 155
Answer key Chapter 11..156

Common Core Test Review.. 159
Common Core Practice Test 1 163
Common Core Practice Test 2 181

Answers and Explanations... 197
Answer Key...199
Practice Test 1 201
Practice Test 2 209

Chapter 1:
Whole Numbers

Add and Subtract Integers

Find the sum or difference.

1) $(+168) + (+76) =$

2) $(+65) + (-32) =$

3) $217 - 69 =$

4) $(-203) + 179 =$

5) $(-45) + 501 =$

6) $182 + (-265) =$

7) $(-9) + 20 =$

8) $360 - 200 =$

9) $(-10) - (-38) =$

10) $(-67) + (-96) =$

11) $(-143) - 234 =$

12) $1250 - (-346) =$

13) $3 + (-12) + (-20) + (-17) =$

14) $(-28) + (-19) + 31 + 16 =$

15) $(-7) - 11 + 27 - 19 =$

16) $6 + (-20) + (-35 - 24) =$

17) $(+24) + (+32) + (-47) =$

18) $(-35) + (-26) =$

19) $-12 - 17 - 16 - 23 =$

20) $7 + (-21) =$

21) $107 - 80 - 73 - (-38) =$

22) $(20) - (-8) =$

23) $(3) - (5) - (-14) =$

24) $(20) - (6) - (-20) =$

Multiplication and Division

Calculate.

1) $340 \times 8 =$

2) $180 \times 30 =$

3) $(-3) \times 7 \times (-4) =$

4) $-3 \times (-6) \times (-6) =$

5) $12 \times (-12) =$

6) $30 \times (-6) =$

7) $6 \times (-1) \times 5 =$

8) $(-600) \times (-50) =$

9) $(-10) \times (-10) \times 2 =$

10) $165 \times 5 =$

11) $160 \times 80 =$

12) $312 \div 12 =$

13) $(-2,475) \div 3 =$

14) $(-32) \div (-8) =$

15) $384 \div (-24) =$

16) $4,500 \div 36 =$

17) $(-84) \div 2 =$

18) $9,588 \div 6 =$

19) $900 \div (-25) =$

20) $1,680 \div 2 =$

21) $(-81) \div 3 =$

22) $(-1,000) \div (-10) =$

23) $0 \div 250 =$

24) $(-680) \div 4 =$

25) $7,704 \div 856 =$

26) $(-3,150) \div 5 =$

27) $7,268 \div 2 =$

28) $(-48) \div (-4)$

Absolute Value

Simplify each equation below.

1) $|-30| =$

2) $-10 + |-30| + 28 =$

3) $|-48| - |-20| + 12 =$

4) $|-9 + 5 - 3| + |3 + 3| =$

5) $2|2 - 14| + 10 =$

6) $|-6| + |-20| =$

7) $|-36 + 20| + 10 - 9 =$

8) $|-10| - |-23| - 5 =$

9) $|-20| - |-10| + 3 =$

10) $|20| - 28 + |-10| =$

11) $\frac{4|3-6|}{2} =$

12) $|-20 + 9| =$

13) $|-20| \times |5| + 5 =$

14) $|-6| + |-36| + 9 - 3 =$

15) $|-20| + |-20| - 40 =$

16) $13 + |-34 + 15| + |-10| =$

17) $28 - |-63| + 10 =$

18) $\frac{|120|}{|4|} + 6 =$

19) $|-9 + 12| + |32 - 15| + 6 =$

20) $|-20 + 15| + |-5| + 3 =$

21) $\frac{|-32|}{8} \times |-6| =$

22) $\frac{4|4 \times 6|}{2} \times \frac{|-16|}{4} =$

23) $\frac{|2 \times 6|}{12} \times 6 =$

24) $|-10 + 2| \times \frac{|-3 \times 5|}{3} =$

25) $|-100 + 8| - 5 + 5 =$

26) $|-50 + 40| - 10 =$

Ordering Integers and Numbers

Order each set of integers from least to greatest.

1) $7, -8, -5, -2, 3$ ___, ___, ___, ___, ___, ___

2) $-3, -16, 4, 10, 9$ ___, ___, ___, ___, ___, ___

3) $18, -18, -19, 25, -20$ ___, ___, ___, ___, ___, ___

4) $-9, -35, 15, -7, 42$ ___, ___, ___, ___, ___, ___

5) $47, -52, 28, -55, 34$ ___, ___, ___, ___, ___, ___

6) $88, 36, -29, 67, -44$ ___, ___, ___, ___, ___, ___

Order each set of integers from greatest to least.

7) $12, 18, -10, -12, -4$ ___, ___, ___, ___, ___, ___

8) $29, 36, -14 -26, 69$ ___, ___, ___, ___, ___, ___

9) $75, -26, -18, 47, -7$ ___, ___, ___, ___, ___, ___

10) $58, 72, -16, -12, 94$ ___, ___, ___, ___, ___, ___

11) $-7, 99, -15, -48, 64$ ___, ___, ___, ___, ___, ___

12) $-80, -45, -40, 18, 29$ ___, ___, ___, ___, ___, ___

Order of Operations

Evaluate each expression.

1) $5 + (4 \times 3) =$

2) $12 - (3 \times 5) =$

3) $(16 \times 3) + 10 =$

4) $(15 - 5) - (6 \times 3) =$

5) $22 + (16 \div 2) =$

6) $(16 \times 5) \div 5 =$

7) $(84 \div 4) \times (-2) =$

8) $(9 \times 5) + (35 - 12) =$

9) $60 + (2 \times 2) + 8 =$

10) $(30 \times 5) \div (2 + 1) =$

11) $(-8) + (10 \times 4) + 13 =$

12) $(7 \times 6) - (32 \div 4) =$

13) $(9 \times 8 \div 3) - (10 + 11) =$

14) $(12 + 8 - 15) \times 6 - 3 =$

15) $(30 - 12 + 40) \times (95 \div 5) =$

16) $22 + (20 - (32 \div 2)) =$

17) $(6 + 9 - 5 - 8) + (18 \div 2) =$

18) $(85 - 10) + (10 - 15 + 9) =$

19) $(10 \times 2) + (12 \times 5) - 12 =$

20) $12 + 8 - (32 \times 4) + 30 =$

Factoring

Factor, write prime if prime.

1) 12

2) 26

3) 32

4) 48

5) 60

6) 64

7) 35

8) 30

9) 56

10) 75

11) 25

12) 18

13) 49

14) 15

15) 42

16) 124

17) 56

18) 40

19) 75

20) 20

21) 96

22) 27

23) 72

24) 50

25) 24

26) 88

27) 68

28) 124

Great Common Factor (GCF)

Find the GCF of the numbers.

1) 8, 12

2) 48, 32

3) 42, 18

4) 10, 15

5) 18, 24

6) 16, 12

7) 80, 45

8) 100, 75

9) 64, 8

10) 36, 72

11) 93, 62

12) 15, 90

13) 60, 30

14) 36, 28

15) 18, 45

16) 35, 42

17) 12, 20

18) 90, 120, 20

19) 49, 144

20) 16, 28

21) 14, 8, 21

22) 4, 16, 20

23) 14, 49, 7

24) 21, 12

Least Common Multiple (LCM)

Find the LCM of each.

1) 6, 9

2) 30, 24

3) 8, 4, 6

4) 15, 12

5) 30, 5, 40

6) 45, 15

7) 15, 10, 8

8) 3, 4

9) 10, 20, 25

10) 64, 44

11) 24, 36

12) 108, 64

13) 20, 10, 40

14) 12, 20

15) 45, 9, 3

16) 27, 63

17) 42, 12

18) 20, 45

19) 25, 15

20) 14, 32

21) 16, 18

22) 9, 17

23) 32, 18

24) 16, 12

Common Core Math Practice Grade 7

Divisibility Rule

Apply the divisibility rules to find the factors of each number.

1) 12	2, 3, 4, 5, 6, 9, 10		13) 18	2, 3, 4, 5, 6, 9, 10
2) 326	2, 3, 4, 5, 6, 9, 10		14) 405	2, 3, 4, 5, 6, 9, 10
3) 748	2, 3, 4, 5, 6, 9, 10		15) 945	2, 3, 4, 5, 6, 9, 10
4) 81	2, 3, 4, 5, 6, 9, 10		16) 186	2, 3, 4, 5, 6, 9, 10
5) 891	2, 3, 4, 5, 6, 9, 10		17) 640	2, 3, 4, 5, 6, 9, 10
6) 345	2, 3, 4, 5, 6, 9, 10		18) 150	2, 3, 4, 5, 6, 9, 10
7) 75	2, 3, 4, 5, 6, 9, 10		19) 350	2, 3, 4, 5, 6, 9, 10
8) 450	2, 3, 4, 5, 6, 9, 10		20) 4,520	2, 3, 4, 5, 6, 9, 10
9) 1,325	2, 3, 4, 5, 6, 9, 10		21) 990	2, 3, 4, 5, 6, 9, 10
10) 78	2, 3, 4, 5, 6, 9, 10		22) 368	2, 3, 4, 5, 6, 9, 10
11) 772	2, 3, 4, 5, 6, 9, 10		23) 208	2, 3, 4, 5, 6, 9, 10
12) 162	2, 3, 4, 5, 6, 9, 10		24) 500	2, 3, 4, 5, 6, 9, 10

WWW.MathNotion.com

Common Core Math Practice Grade 7

Answer key Chapter 1

Add and Subtract Integers

1) 244
2) 33
3) 148
4) −24
5) 456
6) −83
7) 11
8) 160
9) 28
10) −163
11) 377
12) 1,596
13) −46
14) 0
15) −10
16) −73
17) 9
18) −61
19) −68
20) −14
21) −8
22) 28
23) 12
24) 34

Multiplication and Division

1) 2,720
2) 5,400
3) 84
4) −108
5) −144
6) −180
7) −30
8) 30,000
9) 200
10) 825
11) 12,800
12) 26
13) −825
14) 4
15) −16
16) 125
17) −42
18) 1,598
19) −36
20) 840
21) −27
22) 100
23) 0
24) −170
25) 9
26) −630
27) 3,634
28) 12

Absolute Value

1) 30
2) 48
3) 40
4) 13
5) 34
6) 26
7) 17
8) −18
9) 13
10) 2
11) 6
12) 11
13) 105
14) 48
15) 0
16) 42
17) −25
18) 36
19) 26
20) 13
21) 24
22) 192
23) 6
24) 40

Common Core Math Practice Grade 7

25) 92   26) 0

Ordering Integers and Numbers

1) −8, −5, −2, 3, 7
2) −16, −3, 4, 9, 10
3) −20, −19, −18, 18, 25
4) −35, −9, −7, 15, 42
5) −55, −52, 28, 34, 47
6) −44, −29, 36, 67, 88
7) 18, 12, −4, −10, −12
8) 69, 36, 29, −14, −26
9) 75, 47, −7, −18, −26
10) 94, 72, 58, −12, −16
11) 99, 64, −7, −15, −48
12) 29, 18, −40, −45, −80

Order of Operations

1) 17
2) −3
3) 58
4) −8
5) 30
6) 16
7) −42
8) 68
9) 72
10) 50
11) 45
12) 34
13) 3
14) 27
15) 1,102
16) 26
17) 11
18) 79
19) 68
20) −78

Factoring

1) 1, 2, 3, 4, 6, 12
2) 1, 2, 13, 26
3) 1, 2, 4, 8, 16, 32
4) 1, 2, 3, 4, 6, 8, 12, 16, 24, 48
5) 1, 2, 3, 4, 5, 6, 10, 12, 15, 20, 30, 60
6) 1, 2, 4, 8, 16, 32, 64
7) 1, 5, 7, 35
8) 1, 2, 3, 5, 6, 10, 15, 30
9) 1, 2, 4, 7, 8, 14, 28, 56
10) 1, 3, 5, 15, 25, 75
11) 1, 5, 25
12) 1, 2, 3, 6, 9, 18
13) 1, 7, 49
14) 1, 3, 5, 15
15) 1, 2, 3, 6, 7, 14, 21, 42
16) 1, 2, 4, 31, 62, 124
17) 1, 2, 4, 7, 8, 14, 28, 56
18) 1, 2, 4, 5, 8, 10, 20, 40
19) 1, 3, 5, 15, 25, 75
20) 1, 2, 4, 5, 10, 20
21) 1, 2, 3, 4, 6, 8, 12, 24, 32, 48, 96
22) 1, 3, 9, 27
23) 1, 2, 3, 4, 6, 8, 9, 12, 18, 24, 36, 72
24) 1, 2, 5, 10, 25, 50
25) 1, 2, 3, 4, 6, 8, 12, 24
26) 1, 2, 4, 8, 11, 22, 44, 88
27) 1, 2, 4, 17, 34, 68
28) 1, 2, 4, 31, 62, 124

Great Common Factor (GCF)

1) 4
2) 16
3) 6
4) 5
5) 6
6) 4

7) 5	13) 30	19) 1
8) 25	14) 4	20) 4
9) 8	15) 9	21) 1
10) 36	16) 7	22) 4
11) 31	17) 4	23) 7
12) 15	18) 10	24) 3

Least Common Multiple (LCM)

1) 18	9) 100	17) 84
2) 120	10) 704	18) 180
3) 24	11) 72	19) 75
4) 60	12) 1,728	20) 224
5) 120	13) 40	21) 144
6) 45	14) 60	22) 153
7) 120	15) 45	23) 288
8) 12	16) 189	24) 48

Divisibility Rule

1) 12	$\underline{2}$, $\underline{3}$, $\underline{4}$, 5, $\underline{6}$, 9, 10	13) 18	$\underline{2}$, $\underline{3}$, 4, 5, $\underline{6}$, $\underline{9}$, 10
2) 326	$\underline{2}$, 3, 4, 5, 6, 9, 10	14) 405	2, $\underline{3}$, 4, $\underline{5}$, 6, $\underline{9}$, 10
3) 748	$\underline{2}$, 3, $\underline{4}$, 5, 6, 9, 10	15) 945	2, $\underline{3}$, 4, $\underline{5}$, 6, $\underline{9}$, 10
4) 81	2, $\underline{3}$, 4, 5, 6, $\underline{9}$, 10	16) 186	$\underline{2}$, $\underline{3}$, 4, 5, $\underline{6}$, 9, 10
5) 891	2, $\underline{3}$, 4, 5, 6, $\underline{9}$, 10	17) 640	$\underline{2}$, 3, $\underline{4}$, $\underline{5}$, 6, 9, $\underline{10}$
6) 345	2, $\underline{3}$, 4, $\underline{5}$, 6, 9, 10	18) 150	$\underline{2}$, $\underline{3}$, 4, $\underline{5}$, 6, 9, $\underline{10}$
7) 75	2, $\underline{3}$, 4, $\underline{5}$, 6, 9, 10	19) 350	$\underline{2}$, 3, 4, $\underline{5}$, 6, 9, $\underline{10}$
8) 450	$\underline{2}$, $\underline{3}$, 4, $\underline{5}$, $\underline{6}$, $\underline{9}$, $\underline{10}$	20) 4,520	$\underline{2}$, 3, $\underline{4}$, $\underline{5}$, 6, 9, $\underline{10}$
9) 1,325	2, 3, 4, $\underline{5}$, 6, 9, 10	21) 990	$\underline{2}$, $\underline{3}$, 4, $\underline{5}$, $\underline{6}$, $\underline{9}$, $\underline{10}$
10) 78	$\underline{2}$, $\underline{3}$, 4, 5, $\underline{6}$, 9, 10	22) 368	$\underline{2}$, 3, $\underline{4}$, 5, 6, 9, 10
11) 772	$\underline{2}$, 3, $\underline{4}$, 5, 6, 9, 10	23) 208	$\underline{2}$, 3, $\underline{4}$, 5, 6, 9, 10
12) 162	$\underline{2}$, $\underline{3}$, 4, 5, $\underline{6}$, $\underline{9}$, 10	24) 500	$\underline{2}$, 3, $\underline{4}$, $\underline{5}$, 6, 9, $\underline{10}$

Chapter 2:
Fractions

Adding Fractions – Like Denominator

Find each sum.

1) $\frac{1}{4} + \frac{2}{4} =$

2) $\frac{2}{5} + \frac{1}{5} =$

3) $\frac{1}{8} + \frac{2}{8} =$

4) $\frac{4}{11} + \frac{1}{11} =$

5) $\frac{4}{21} + \frac{1}{21} =$

6) $\frac{5}{49} + \frac{6}{49} =$

7) $\frac{2}{7} + \frac{11}{7} =$

8) $\frac{1}{15} + \frac{3}{15} =$

9) $\frac{3}{19} + \frac{6}{19} =$

10) $\frac{1}{13} + \frac{1}{13} =$

11) $\frac{1}{5} + \frac{1}{5} =$

12) $\frac{4}{17} + \frac{6}{17} =$

13) $\frac{2}{20} + \frac{17}{20} =$

14) $\frac{4}{25} + \frac{7}{25} =$

15) $\frac{6}{14} + \frac{3}{14} =$

16) $\frac{12}{30} + \frac{5}{30} =$

17) $\frac{1}{9} + \frac{1}{9} =$

18) $\frac{29}{5} + \frac{3}{5} =$

19) $\frac{18}{6} + \frac{5}{6} =$

20) $\frac{25}{37} + \frac{11}{37} =$

Adding Fractions – Unlike Denominator

Add the fractions and simplify the answers.

1) $\frac{1}{3} + \frac{1}{2} =$

2) $\frac{2}{7} + \frac{2}{3} =$

3) $\frac{3}{6} + \frac{1}{5} =$

4) $\frac{5}{13} + \frac{2}{4} =$

5) $\frac{3}{15} + \frac{2}{5} =$

6) $\frac{16}{56} + \frac{3}{16} =$

7) $\frac{3}{7} + \frac{2}{5} =$

8) $\frac{4}{12} + \frac{2}{5} =$

9) $\frac{6}{13} + \frac{3}{7} =$

10) $\frac{3}{8} + \frac{2}{5} =$

11) $\frac{1}{16} + \frac{4}{6} =$

12) $\frac{5}{24} + \frac{2}{3} =$

13) $\frac{3}{36} + \frac{5}{4} =$

14) $\frac{1}{25} + \frac{2}{5} =$

15) $\frac{7}{49} + \frac{3}{7} =$

16) $\frac{7}{12} + \frac{5}{6} =$

17) $\frac{3}{9} + \frac{2}{5} =$

18) $\frac{3}{45} + \frac{1}{5} =$

19) $\frac{3}{18} + \frac{7}{4} =$

20) $\frac{3}{10} + \frac{1}{4} =$

21) $\frac{3}{64} + \frac{1}{8} =$

22) $\frac{6}{14} + \frac{1}{3} =$

23) $\frac{2}{81} + \frac{1}{3} =$

24) $\frac{6}{15} + \frac{1}{3} =$

Subtracting Fractions – Like Denominator

Find the difference.

1) $\dfrac{5}{3} - \dfrac{2}{3} =$

2) $\dfrac{5}{8} - \dfrac{3}{8} =$

3) $\dfrac{11}{14} - \dfrac{8}{14} =$

4) $\dfrac{13}{3} - \dfrac{7}{3} =$

5) $\dfrac{15}{17} - \dfrac{13}{17} =$

6) $\dfrac{18}{33} - \dfrac{10}{33} =$

7) $\dfrac{8}{25} - \dfrac{2}{25} =$

8) $\dfrac{17}{27} - \dfrac{2}{27} =$

9) $\dfrac{7}{10} - \dfrac{3}{10} =$

10) $\dfrac{24}{35} - \dfrac{4}{35} =$

11) $\dfrac{11}{5} - \dfrac{3}{5} =$

12) $\dfrac{28}{38} - \dfrac{18}{38} =$

13) $\dfrac{5}{6} - \dfrac{1}{6} =$

14) $\dfrac{22}{43} - \dfrac{11}{43} =$

15) $\dfrac{4}{7} - \dfrac{3}{7} =$

16) $\dfrac{18}{29} - \dfrac{15}{29} =$

17) $\dfrac{4}{5} - \dfrac{3}{5} =$

18) $\dfrac{42}{53} - \dfrac{38}{53} =$

19) $\dfrac{8}{31} - \dfrac{3}{31} =$

20) $\dfrac{32}{39} - \dfrac{30}{39} =$

21) $\dfrac{9}{26} - \dfrac{5}{26} =$

22) $\dfrac{31}{46} - \dfrac{27}{46} =$

23) $\dfrac{25}{48} - \dfrac{19}{48} =$

24) $\dfrac{39}{65} - \dfrac{27}{65} =$

Subtracting Fractions – Unlike Denominator

Solve each problem.

1) $\dfrac{1}{2} - \dfrac{1}{3} =$

2) $\dfrac{5}{8} - \dfrac{2}{5} =$

3) $\dfrac{5}{6} - \dfrac{2}{7} =$

4) $\dfrac{3}{5} - \dfrac{1}{10} =$

5) $\dfrac{3}{5} - \dfrac{5}{12} =$

6) $\dfrac{5}{8} - \dfrac{5}{16} =$

7) $\dfrac{2}{25} - \dfrac{1}{15} =$

8) $\dfrac{3}{4} - \dfrac{13}{18} =$

9) $\dfrac{8}{5} - \dfrac{7}{6} =$

10) $\dfrac{5}{6} - \dfrac{2}{24} =$

11) $\dfrac{3}{4} - \dfrac{5}{36} =$

12) $\dfrac{1}{5} - \dfrac{2}{25} =$

13) $\dfrac{7}{6} - \dfrac{3}{18} =$

14) $\dfrac{7}{6} - \dfrac{5}{12} =$

15) $\dfrac{3}{5} - \dfrac{2}{9} =$

16) $\dfrac{3}{5} - \dfrac{1}{45} =$

17) $\dfrac{5}{32} - \dfrac{5}{48} =$

18) $\dfrac{2}{3} - \dfrac{2}{7} =$

19) $\dfrac{3}{5} - \dfrac{1}{6} =$

20) $\dfrac{3}{4} - \dfrac{5}{13} =$

Common Core Math Practice Grade 7

Converting Mix Numbers

Convert the following mixed numbers into improper fractions.

1) $2\frac{3}{4} =$

2) $4\frac{12}{65} =$

3) $9\frac{3}{7} =$

4) $3\frac{5}{6} =$

5) $6\frac{6}{7} =$

6) $2\frac{10}{24} =$

7) $6\frac{7}{12} =$

8) $2\frac{12}{13} =$

9) $2\frac{12}{10} =$

10) $8\frac{6}{7} =$

11) $6\frac{1}{2} =$

12) $5\frac{14}{16} =$

13) $4\frac{8}{7} =$

14) $2\frac{9}{12} =$

15) $8\frac{3}{5} =$

16) $3\frac{4}{12} =$

17) $6\frac{3}{7} =$

18) $2\frac{1}{15} =$

19) $3\frac{7}{15} =$

20) $4\frac{3}{4} =$

21) $3\frac{5}{9} =$

22) $2\frac{11}{5} =$

23) $5\frac{13}{3} =$

24) $11\frac{7}{13} =$

WWW.MathNotion.com

Converting improper Fractions

Convert the following improper fractions into mixed numbers

1) $\frac{67}{12} =$

2) $\frac{75}{63} =$

3) $\frac{19}{15} =$

4) $\frac{58}{45} =$

5) $\frac{85}{26} =$

6) $\frac{271}{52} =$

7) $\frac{84}{63} =$

8) $\frac{41}{5} =$

9) $\frac{16}{15} =$

10) $\frac{11}{2} =$

11) $\frac{35}{4} =$

12) $\frac{120}{95} =$

13) $\frac{120}{54} =$

14) $\frac{28}{8} =$

15) $\frac{83}{11} =$

16) $\frac{31}{3} =$

17) $\frac{101}{8} =$

18) $\frac{51}{48} =$

19) $\frac{28}{9} =$

20) $\frac{8}{7} =$

21) $\frac{7}{2} =$

22) $\frac{43}{10} =$

23) $\frac{32}{24} =$

24) $\frac{78}{7} =$

Addition Mix Numbers

Add the following fractions.

1) $2\frac{1}{3} + 3\frac{1}{3} =$

2) $6\frac{3}{4} + 2\frac{1}{4} =$

3) $1\frac{1}{7} + 2\frac{2}{7} =$

4) $3\frac{1}{6} + 2\frac{3}{2} =$

5) $3\frac{4}{12} + 3\frac{3}{10} =$

6) $4\frac{1}{7} + 1\frac{1}{2} =$

7) $1\frac{2}{21} + 1\frac{2}{24} =$

8) $3\frac{2}{5} + 1\frac{3}{2} =$

9) $2\frac{3}{5} + 2\frac{1}{5} =$

10) $2\frac{4}{5} + 1\frac{3}{5} =$

11) $3\frac{2}{3} + 1\frac{3}{4} =$

12) $4\frac{1}{6} + 1\frac{3}{7} =$

13) $4\frac{1}{2} + 1\frac{3}{2} =$

14) $5\frac{3}{8} + 2\frac{1}{3} =$

15) $2\frac{3}{4} + 3\frac{1}{3} =$

16) $3\frac{1}{4} + 2\frac{3}{5} =$

17) $2\frac{3}{4} + 8\frac{2}{5} =$

18) $1\frac{3}{4} + 1\frac{1}{2} =$

19) $2\frac{3}{4} + 1\frac{1}{7} =$

20) $1\frac{2}{3} + 1\frac{3}{4} =$

21) $3\frac{1}{6} + 2\frac{1}{4} =$

22) $8\frac{2}{5} + 2\frac{3}{4} =$

23) $4\frac{2}{3} + 5\frac{1}{7} =$

24) $2\frac{1}{3} + 3\frac{2}{5} =$

Subtracting Mix Numbers

Subtract the following fractions.

1) $4\frac{1}{2} - 3\frac{1}{2} =$

2) $3\frac{3}{7} - 3\frac{1}{7} =$

3) $6\frac{3}{5} - 5\frac{1}{5} =$

4) $3\frac{1}{3} - 2\frac{1}{2} =$

5) $4\frac{1}{5} - 3\frac{1}{2} =$

6) $9\frac{1}{3} - 5\frac{2}{3} =$

7) $5\frac{5}{10} - 1\frac{6}{10} =$

8) $7\frac{4}{9} - 5\frac{8}{9} =$

9) $6\frac{2}{11} - 5\frac{5}{11} =$

10) $6\frac{2}{5} - 1\frac{1}{5} =$

11) $9\frac{1}{2} - 5\frac{1}{4} =$

12) $2\frac{5}{8} - 1\frac{3}{8} =$

13) $5\frac{3}{58} - 2\frac{5}{6} =$

14) $5\frac{1}{4} - 3\frac{1}{2} =$

15) $17\frac{1}{8} - 12\frac{3}{8} =$

16) $3\frac{3}{5} - 2\frac{1}{5} =$

17) $2\frac{1}{3} - 1\frac{2}{3} =$

18) $2\frac{1}{6} - 1\frac{2}{3} =$

19) $3\frac{2}{6} - 2\frac{1}{2} =$

20) $2\frac{5}{3} - 2\frac{1}{12} =$

21) $2\frac{9}{10} - 1\frac{1}{5} =$

22) $4\frac{2}{5} - 3\frac{1}{11} =$

23) $2\frac{1}{2} - 1\frac{1}{6} =$

24) $2\frac{3}{10} - 1\frac{4}{10} =$

Common Core Math Practice Grade 7

Simplify Fractions

Reduce these fractions to lowest terms

1) $\frac{24}{16} =$

2) $\frac{18}{27} =$

3) $\frac{12}{15} =$

4) $\frac{36}{48} =$

5) $\frac{9}{27} =$

6) $\frac{15}{35} =$

7) $\frac{28}{49} =$

8) $\frac{80}{100} =$

9) $\frac{9}{81} =$

10) $\frac{25}{10} =$

11) $\frac{24}{32} =$

12) $\frac{20}{60} =$

13) $\frac{24}{40} =$

14) $\frac{3}{12} =$

15) $\frac{14}{49} =$

16) $\frac{52}{78} =$

17) $\frac{96}{36} =$

18) $\frac{48}{180} =$

19) $\frac{12}{32} =$

20) $\frac{88}{77} =$

21) $\frac{160}{320} =$

22) $\frac{24}{124} =$

23) $\frac{144}{36} =$

24) $\frac{120}{480} =$

WWW.MathNotion.com

Multiplying Fractions

Find the product.

1) $\frac{2}{7} \times \frac{3}{8} =$

2) $\frac{4}{25} \times \frac{5}{8} =$

3) $\frac{9}{40} \times \frac{10}{27} =$

4) $\frac{6}{13} \times \frac{22}{33} =$

5) $\frac{9}{12} \times \frac{3}{5} =$

6) $\frac{12}{17} \times \frac{5}{3} =$

7) $\frac{5}{6} \times \frac{6}{5} =$

8) $\frac{35}{89} \times 0 =$

9) $\frac{9}{4} \times \frac{12}{5} =$

10) $\frac{10}{18} \times \frac{3}{5} =$

11) $\frac{36}{25} \times \frac{25}{36} =$

12) $\frac{3}{36} \times \frac{6}{27} =$

13) $\frac{15}{7} \times \frac{3}{5} =$

14) $\frac{6}{7} \times \frac{3}{5} =$

15) $\frac{27}{14} \times \frac{7}{3} =$

16) $\frac{12}{17} \times 0 =$

17) $\frac{7}{11} \times \frac{33}{14} =$

18) $\frac{20}{9} \times \frac{3}{5} =$

19) $\frac{9}{16} \times \frac{4}{81} =$

20) $\frac{4}{23} \times \frac{2}{32} =$

21) $\frac{2}{12} \times \frac{3}{16} =$

22) $\frac{25}{8} \times \frac{2}{125} =$

23) $\frac{9}{16} \times \frac{4}{81} =$

24) $\frac{100}{200} \times \frac{400}{800} =$

WWW.MathNotion.com

Multiplying Mixed Number

Multiply. Reduce to lowest terms.

1) $1\frac{2}{3} \times 1\frac{1}{4} =$

2) $1\frac{2}{5} \times 1\frac{3}{2} =$

3) $1\frac{2}{3} \times 3\frac{1}{8} =$

4) $2\frac{1}{8} \times 1\frac{3}{5} =$

5) $2\frac{2}{3} \times 3\frac{1}{3} =$

6) $2\frac{1}{3} \times 1\frac{2}{3} =$

7) $1\frac{3}{4} \times 2\frac{1}{2} =$

8) $3\frac{2}{3} \times 2\frac{1}{3} =$

9) $2\frac{2}{3} \times 2\frac{1}{2} =$

10) $1\frac{1}{3} \times 1\frac{1}{2} =$

11) $2\frac{3}{4} \times 2\frac{2}{3} =$

12) $3\frac{2}{5} \times 2\frac{4}{7} =$

13) $1\frac{3}{4} \times 2\frac{1}{2} =$

14) $1\frac{1}{2} \times 3\frac{1}{7} =$

15) $1\frac{1}{2} \times 2\frac{1}{5} =$

16) $1\frac{2}{7} \times 2\frac{2}{3} =$

17) $1\frac{2}{3} \times 2\frac{1}{5} =$

18) $1\frac{2}{3} \times 3\frac{2}{5} =$

19) $1\frac{3}{4} \times 2\frac{1}{7} =$

20) $1\frac{1}{3} \times 3\frac{2}{5} =$

21) $1\frac{1}{2} \times 2\frac{1}{6} =$

22) $1\frac{1}{9} \times 1\frac{1}{7} =$

Dividing Fractions

Divide these fractions.

1) $0 \div \frac{1}{5} =$

2) $\frac{6}{12} \div 6 =$

3) $\frac{8}{11} \div \frac{3}{4} =$

4) $\frac{14}{64} \div \frac{2}{8} =$

5) $\frac{3}{19} \div \frac{9}{19} =$

6) $\frac{3}{12} \div \frac{15}{36} =$

7) $9 \div \frac{1}{5} =$

8) $\frac{15}{14} \div \frac{3}{7} =$

9) $\frac{6}{15} \div \frac{1}{14} =$

10) $\frac{2}{13} \div \frac{6}{5} =$

11) $\frac{5}{11} \div \frac{3}{10} =$

12) $\frac{15}{28} \div \frac{3}{7} =$

13) $\frac{7}{16} \div \frac{7}{4} =$

14) $\frac{6}{14} \div \frac{30}{7} =$

15) $\frac{8}{23} \div \frac{2}{23} =$

16) $\frac{9}{32} \div \frac{81}{4} =$

17) $\frac{5}{3} \div \frac{10}{27} =$

18) $8 \div \frac{1}{3} =$

19) $\frac{72}{32} \div \frac{3}{9} =$

20) $\frac{2}{30} \div \frac{8}{5} =$

21) $\frac{2}{9} \div \frac{6}{15} =$

22) $\frac{7}{21} \div \frac{3}{4} =$

Dividing Mixed Number

Divide the following mixed numbers. Cancel and simplify when possible.

1) $2\frac{1}{3} \div 2\frac{1}{2} =$

2) $3\frac{1}{8} \div 2\frac{2}{4} =$

3) $3\frac{1}{2} \div 2\frac{3}{5} =$

4) $2\frac{1}{7} \div 2\frac{1}{2} =$

5) $4\frac{1}{5} \div 2\frac{1}{3} =$

6) $2\frac{5}{9} \div 1\frac{2}{5} =$

7) $2\frac{2}{9} \div 1\frac{1}{2} =$

8) $3\frac{1}{7} \div 2\frac{1}{7} =$

9) $2\frac{1}{9} \div 2\frac{1}{2} =$

10) $3\frac{1}{6} \div 2\frac{2}{3} =$

11) $1\frac{2}{3} \div 5\frac{1}{3} =$

12) $3\frac{1}{9} \div 2\frac{2}{3} =$

13) $3\frac{1}{7} \div 1\frac{1}{11} =$

14) $9\frac{4}{7} \div 4\frac{1}{2} =$

15) $3\frac{3}{4} \div 2\frac{1}{2} =$

16) $2\frac{1}{3} \div 3\frac{2}{5} =$

17) $8\frac{3}{4} \div 2\frac{5}{8} =$

18) $3\frac{1}{3} \div 2\frac{3}{5} =$

19) $3\frac{2}{5} \div 2\frac{1}{2} =$

20) $5\frac{3}{8} \div 2\frac{1}{6} =$

21) $6\frac{1}{2} \div 2\frac{1}{4} =$

22) $4\frac{1}{5} \div 2\frac{1}{7} =$

23) $3\frac{1}{5} \div 2\frac{1}{5} =$

24) $2\frac{1}{7} \div 2\frac{1}{5} =$

WWW.MathNotion.com

Comparing Fractions

Compare the fractions, and write >, < or =

1) $\dfrac{15}{4}$ ___ $\dfrac{31}{12}$

2) $\dfrac{34}{5}$ ___ $\dfrac{1}{4}$

3) $\dfrac{3}{6}$ ___ $\dfrac{7}{5}$

4) $\dfrac{28}{7}$ ___ $\dfrac{14}{5}$

5) $\dfrac{1}{6}$ ___ $\dfrac{3}{5}$

6) $\dfrac{11}{7}$ ___ $\dfrac{15}{9}$

7) $\dfrac{6}{10}$ ___ $\dfrac{4}{7}$

8) $\dfrac{21}{12}$ ___ $\dfrac{23}{6}$

9) $2\dfrac{1}{10}$ ___ $5\dfrac{1}{2}$

10) $4\dfrac{1}{7}$ ___ $2\dfrac{1}{6}$

11) $2\dfrac{1}{3}$ ___ $2\dfrac{1}{4}$

12) $8\dfrac{6}{7}$ ___ $8\dfrac{2}{3}$

13) $1\dfrac{3}{7}$ ___ $2\dfrac{5}{3}$

14) $\dfrac{1}{13}$ ___ $\dfrac{4}{7}$

15) $\dfrac{41}{65}$ ___ $\dfrac{17}{43}$

16) $\dfrac{65}{200}$ ___ $\dfrac{65}{92}$

17) $12\dfrac{1}{2}$ ___ $12\dfrac{1}{7}$

18) $\dfrac{1}{2}$ ___ $\dfrac{1}{4}$

19) $\dfrac{1}{9}$ ___ $\dfrac{1}{15}$

20) $\dfrac{8}{14}$ ___ $\dfrac{6}{10}$

21) $\dfrac{5}{25}$ ___ $\dfrac{8}{56}$

22) $\dfrac{6}{7}$ ___ $\dfrac{3}{7}$

23) $1\dfrac{38}{32}$ ___ $2\dfrac{3}{16}$

24) $4\dfrac{18}{5}$ ___ $5\dfrac{4}{3}$

Answer key Chapter 2

Adding Fractions – Like Denominator

1) $\frac{3}{4}$
2) $\frac{3}{5}$
3) $\frac{3}{8}$
4) $\frac{5}{11}$
5) $\frac{5}{21}$
6) $\frac{11}{49}$
7) $\frac{13}{7}$
8) $\frac{4}{15}$
9) $\frac{9}{19}$
10) $\frac{2}{13}$
11) $\frac{2}{5}$
12) $\frac{10}{17}$
13) $\frac{19}{20}$
14) $\frac{11}{25}$
15) $\frac{9}{14}$
16) $\frac{17}{30}$
17) $\frac{2}{9}$
18) $\frac{32}{5}$
19) $\frac{23}{6}$
20) $\frac{36}{37}$

Adding Fractions – Unlike Denominator

1) $\frac{5}{6}$
2) $\frac{20}{21}$
3) $\frac{7}{10}$
4) $\frac{23}{26}$
5) $\frac{3}{5}$
6) $\frac{53}{112}$
7) $\frac{29}{35}$
8) $\frac{11}{15}$
9) $\frac{81}{91}$
10) $\frac{31}{40}$
11) $\frac{35}{48}$
12) $\frac{7}{8}$
13) $\frac{4}{3}$
14) $\frac{11}{25}$
15) $\frac{4}{7}$
16) $\frac{17}{12}$
17) $\frac{11}{15}$
18) $\frac{4}{15}$
19) $\frac{23}{12}$
20) $\frac{11}{20}$
21) $\frac{11}{64}$
22) $\frac{16}{21}$
23) $\frac{29}{81}$
24) $\frac{11}{15}$

Subtracting Fractions – Like Denominator

1) 1
2) $\frac{1}{4}$
3) $\frac{3}{14}$
4) 2
5) $\frac{2}{17}$
6) $\frac{8}{33}$
7) $\frac{6}{25}$
8) $\frac{5}{9}$
9) $\frac{2}{5}$
10) $\frac{4}{7}$
11) $\frac{8}{5}$
12) $\frac{5}{19}$
13) $\frac{2}{3}$
14) $\frac{11}{43}$
15) $\frac{1}{7}$
16) $\frac{3}{29}$
17) $\frac{1}{5}$
18) $\frac{4}{53}$

Common Core Math Practice Grade 7

19) $\frac{5}{31}$ 21) $\frac{2}{13}$ 23) $\frac{1}{8}$

20) $\frac{2}{39}$ 22) $\frac{2}{23}$ 24) $\frac{12}{65}$

Subtracting Fractions – Unlike Denominator

1) $\frac{1}{6}$ 8) $\frac{1}{36}$ 15) $\frac{17}{45}$

2) $\frac{9}{40}$ 9) $\frac{13}{30}$ 16) $\frac{26}{45}$

3) $\frac{23}{42}$ 10) $\frac{3}{4}$ 17) $\frac{5}{96}$

4) $\frac{1}{2}$ 11) $\frac{11}{18}$ 18) $\frac{8}{21}$

5) $\frac{11}{60}$ 12) $\frac{3}{25}$ 19) $\frac{13}{30}$

6) $\frac{5}{16}$ 13) 1 20) $\frac{19}{52}$

7) $\frac{1}{75}$ 14) $\frac{3}{4}$

Converting Mix Numbers

1) $\frac{11}{4}$ 9) $\frac{32}{10}$ 17) $\frac{45}{7}$

2) $\frac{272}{65}$ 10) $\frac{62}{7}$ 18) $\frac{31}{15}$

3) $\frac{66}{7}$ 11) $\frac{13}{2}$ 19) $\frac{52}{15}$

4) $\frac{23}{6}$ 12) $\frac{94}{16}$ 20) $\frac{19}{4}$

5) $\frac{48}{7}$ 13) $\frac{36}{7}$ 21) $\frac{32}{9}$

6) $\frac{58}{24}$ 14) $\frac{33}{12}$ 22) $\frac{21}{5}$

7) $\frac{79}{12}$ 15) $\frac{43}{5}$ 23) $\frac{28}{3}$

8) $\frac{38}{13}$ 16) $\frac{40}{12}$ 24) $\frac{150}{13}$

Converting improper Fractions

1) $5\frac{7}{12}$ 6) $5\frac{11}{52}$ 11) $8\frac{3}{4}$

2) $1\frac{21}{63}$ 7) $1\frac{21}{63}$ 12) $1\frac{25}{95}$

3) $1\frac{4}{15}$ 8) $8\frac{1}{5}$ 13) $2\frac{12}{54}$

4) $1\frac{13}{45}$ 9) $1\frac{1}{15}$ 14) $3\frac{4}{8}$

5) $3\frac{7}{26}$ 10) $5\frac{1}{2}$ 15) $7\frac{6}{11}$

Common Core Math Practice Grade 7

16) $10\frac{1}{3}$
17) $12\frac{5}{8}$
18) $1\frac{1}{16}$

19) $3\frac{1}{9}$
20) $1\frac{1}{7}$
21) $3\frac{1}{2}$

22) $4\frac{3}{10}$
23) $1\frac{1}{3}$
24) $11\frac{1}{7}$

Adding Mix Numbers

1) $5\frac{2}{3}$
2) 9
3) $3\frac{3}{7}$
4) $6\frac{2}{3}$
5) $6\frac{19}{30}$
6) $5\frac{9}{14}$
7) $2\frac{5}{28}$
8) $5\frac{9}{10}$

9) $4\frac{4}{5}$
10) $4\frac{2}{5}$
11) $5\frac{5}{12}$
12) $5\frac{25}{42}$
13) 7
14) $7\frac{17}{24}$
15) $6\frac{1}{12}$
16) $5\frac{17}{20}$

17) $11\frac{3}{20}$
18) $3\frac{1}{4}$
19) $3\frac{25}{28}$
20) $3\frac{5}{12}$
21) $5\frac{5}{12}$
22) $11\frac{3}{20}$
23) $9\frac{17}{21}$
24) $5\frac{11}{15}$

Subtracting Mix Numbers

1) 1
2) $\frac{2}{7}$
3) $1\frac{2}{5}$
4) $\frac{5}{6}$
5) $\frac{7}{10}$
6) $3\frac{2}{3}$
7) $3\frac{9}{10}$
8) $1\frac{5}{9}$

9) $\frac{8}{11}$
10) $5\frac{1}{5}$
11) $4\frac{1}{4}$
12) $1\frac{1}{4}$
13) $2\frac{19}{87}$
14) $1\frac{3}{4}$
15) $4\frac{3}{4}$
16) $1\frac{2}{5}$

17) $\frac{2}{3}$
18) $\frac{1}{2}$
19) $\frac{5}{6}$
20) $1\frac{7}{12}$
21) $1\frac{7}{10}$
22) $1\frac{17}{55}$
23) $1\frac{1}{3}$
24) $\frac{9}{10}$

Simplify Fractions

1) $\frac{3}{2}$
2) $\frac{2}{3}$
3) $\frac{4}{5}$

4) $\frac{3}{4}$
5) $\frac{1}{3}$
6) $\frac{3}{7}$

7) $\frac{4}{7}$
8) $\frac{4}{5}$
9) $\frac{1}{9}$

WWW.MathNotion.com

Common Core Math Practice Grade 7

10) $\frac{5}{2}$

11) $\frac{3}{4}$

12) $\frac{1}{3}$

13) $\frac{3}{5}$

14) $\frac{1}{4}$

15) $\frac{2}{7}$

16) $\frac{2}{3}$

17) $\frac{8}{3}$

18) $\frac{4}{15}$

19) $\frac{3}{8}$

20) $\frac{8}{7}$

21) $\frac{1}{2}$

22) $\frac{6}{31}$

23) 4

24) $\frac{1}{4}$

Multiplying Fractions

1) $\frac{3}{28}$

2) $\frac{1}{10}$

3) $\frac{1}{12}$

4) $\frac{4}{13}$

5) $\frac{9}{20}$

6) $\frac{20}{17}$

7) 1

8) 0

9) $\frac{27}{5}$

10) $\frac{1}{3}$

11) 1

12) $\frac{1}{54}$

13) $\frac{9}{7}$

14) $\frac{18}{35}$

15) $\frac{9}{2}$

16) 0

17) $\frac{3}{2}$

18) $\frac{4}{3}$

19) $\frac{1}{36}$

20) $\frac{1}{92}$

21) $\frac{1}{32}$

22) $\frac{1}{20}$

23) $\frac{1}{36}$

24) $\frac{1}{4}$

Multiplying Mixed Number

1) $2\frac{1}{12}$

2) $3\frac{1}{2}$

3) $5\frac{5}{24}$

4) $3\frac{2}{5}$

5) $8\frac{8}{9}$

6) $3\frac{8}{9}$

7) $4\frac{3}{8}$

8) $8\frac{5}{9}$

9) $6\frac{2}{3}$

10) 2

11) $7\frac{1}{3}$

12) $8\frac{26}{35}$

13) $4\frac{3}{8}$

14) $4\frac{5}{7}$

15) $3\frac{3}{10}$

16) $3\frac{3}{7}$

17) $3\frac{2}{3}$

18) $5\frac{2}{3}$

19) $3\frac{3}{4}$

20) $4\frac{8}{15}$

21) $3\frac{1}{4}$

22) $1\frac{17}{63}$

Dividing Fractions

1) 0

Common Core Math Practice Grade 7

2) $\frac{1}{12}$
3) $\frac{32}{33}$
4) $\frac{7}{8}$
5) $\frac{1}{3}$
6) $\frac{3}{5}$
7) 45
8) $\frac{5}{2}$

9) $\frac{28}{5}$
10) $\frac{5}{39}$
11) $\frac{50}{33}$
12) $\frac{5}{4}$
13) $\frac{1}{4}$
14) $\frac{1}{10}$
15) 4

16) $\frac{1}{72}$
17) $\frac{9}{2}$
18) 24
19) $\frac{27}{4}$
20) $\frac{1}{24}$
21) $\frac{5}{9}$
22) $\frac{4}{9}$

Dividing Mixed Number

1) $\frac{14}{15}$
2) $1\frac{1}{4}$
3) $1\frac{9}{26}$
4) $\frac{6}{7}$
5) $1\frac{4}{5}$
6) $1\frac{52}{63}$
7) $1\frac{13}{27}$
8) $1\frac{7}{15}$

9) $\frac{38}{45}$
10) $1\frac{3}{16}$
11) $\frac{5}{16}$
12) $1\frac{1}{6}$
13) $2\frac{37}{42}$
14) $2\frac{8}{63}$
15) $1\frac{1}{2}$
16) $\frac{35}{51}$

17) $3\frac{1}{3}$
18) $1\frac{11}{39}$
19) $1\frac{9}{25}$
20) $2\frac{25}{52}$
21) $2\frac{8}{9}$
22) $1\frac{24}{25}$
23) $1\frac{5}{11}$
24) $\frac{75}{77}$

Comparing Fractions

1) >
2) >
3) <
4) >
5) <
6) <

7) >
8) <
9) <
10) >
11) >
12) >

13) <
14) <
15) >
16) <
17) >
18) >

19) >
20) <
21) >
22) >
23) =
24) >

Chapter 3:
Decimal

Round Decimals

Round each number to the correct place value

1) 0.6<u>4</u> =

2) 2.<u>0</u>4 =

3) 6.<u>6</u>23 =

4) 0.<u>3</u>77 =

5) <u>7</u>.707 =

6) 0.0<u>8</u>9 =

7) 6.<u>2</u>4 =

8) 76.7<u>6</u>0 =

9) 1.6<u>2</u>9 =

10) 10.<u>3</u>858 =

11) 1.<u>0</u>9 =

12) 4.<u>2</u>32 =

13) 3.<u>2</u>43 =

14) 6.0<u>5</u>20 =

15) 6<u>3</u>.69 =

16) 3<u>7</u>.32 =

17) 4<u>1</u>9.078 =

18) 512.<u>6</u>55 =

19) 12.3<u>6</u>2 =

20) 6<u>5</u>.65 =

21) 3.2<u>0</u>89 =

22) 37.<u>0</u>73 =

23) 126.<u>5</u>16 =

24) 0.0<u>1</u>22 =

25) 0.07<u>8</u>5 =

26) 5.0<u>1</u>62 =

27) 23.6<u>1</u>33 =

28) 8.0<u>8</u>20 =

Decimals Addition

Add the following.

1) 25.52 + 52.25

2) 0.93 + 0.07

3) 18.96 + 12.87

4) 56.106 + 3.198

5) 6.960 + 5.87

6) 4.148 + 3.231

7) 72.72 + 12.87

8) 56.24 + 23.47

9) 43.06 + 11.87

10) 7.961 + 12.87

11) 18.148 + 12.231

12) 65.98 + 8.37

13) 28.05 + 7.37

14) 125.32 + 3.32

Decimals Subtraction

Subtract the following

1) 8.97 − 2.82

2) 84.02 − 67.57

3) 0.65 − 0.2

4) 9.784 − 7.2

5) 0.784 − 0.05

6) 84.62 − 23.81

7) 121.26 − 78.97

8) 24.36 − 8.38

9) 52.59 − 37.6

10) 5.872 − 0.297

11) 61.43 − 18.8

12) 17.732 − 4.314

13) 23.502 − 2.817

14) 135.35 − 23.56

Common Core Math Practice Grade 7

Decimals Multiplication

Solve.

1) 2.1 × 2.6

2) 8.7 × 5.9

3) 7.06 × 2.05

4) 67.08 × 10

5) 13.08 × 1000

6) 32.06 × 7.8

7) 26.12 × 12.01

8) 4.06 × 7.05

9) 18.06 × 0.05

10) 21.09 × 9.07

11) 14.3 × 15.7

12) 5.12 × 0.03

13) 8.05 × 0.21

14) 12.12 × 5.03

Decimal Division

Dividing Decimals.

1) $7 \div 1{,}000 =$

2) $3 \div 10 =$

3) $2.6 \div 1{,}000 =$

4) $0.01 \div 100 =$

5) $7 \div 49 =$

6) $2 \div 82 =$

7) $3 \div 48 =$

8) $8 \div 120 =$

9) $8 \div 100 =$

10) $0.8 \div 0.72 =$

11) $0.7 \div 0.07 =$

12) $0.9 \div 0.36 =$

13) $0.5 \div 0.35 =$

14) $0.6 \div 0.06 =$

15) $2.07 \div 10 =$

16) $7.6 \div 100 =$

17) $7.38 \div 1{,}000 =$

18) $15.6 \div 4.5 =$

19) $45.2 \div 5 =$

20) $0.3 \div 0.03 =$

21) $8.05 \div 2.5 =$

22) $0.05 \div 0.20 =$

23) $0.7 \div 4.4 =$

24) $0.08 \div 50 =$

25) $4.16 \div 0.8 =$

26) $0.08 \div 384 =$

Comparing Decimals

Write the Correct Comparison Symbol (>, < or =)

1) 1.15 _____ 2.15

2) 0.4 _____ 0.385

3) 12.5 _____ 12.500

4) 4.05 _____ 4.50

5) 0.511 _____ 0.51

6) 0.623 _____ 0.723

7) 8.76 _____ 8.678

8) 3.0069 _____ 3.069

9) 23.042 _____ 23.034

10) 6.11 _____ 6.08

11) 2.22 _____ 2.222

12) 0.06 _____ 0.55

13) 1.204 _____ 1.25

14) 4.92 _____ 4.0952

15) 0.44 _____ 0.044

16) 17.04 _____ 17.040

17) 0.090 _____ 0.80

18) 20.217 _____ 22.1

19) 0.021 _____ 0.201

20) 21.5 _____ 11.8

21) 3.5 _____ 10.9

22) 0.071 _____ 0.0701

23) 4.021 _____ 0.4021

24) 2.5 _____ 0.255

25) 5.2 _____ 0.255

26) 2.05 _____ 2.0500

27) 6.05 _____ 0.655

28) 1.0501 _____ 1.0510

Convert Fraction to Decimal

Write each as a decimal.

1) $\frac{40}{100} =$

2) $\frac{38}{100} =$

3) $\frac{4}{25} =$

4) $\frac{6}{24} =$

5) $\frac{9}{81} =$

6) $\frac{49}{100} =$

7) $\frac{2}{25} =$

8) $\frac{17}{25} =$

9) $\frac{47}{200} =$

10) $\frac{13}{50} =$

11) $\frac{18}{36} =$

12) $\frac{3}{8} =$

13) $\frac{6}{20} =$

14) $\frac{9}{125} =$

15) $\frac{27}{50} =$

16) $\frac{20}{50} =$

17) $\frac{45}{10} =$

18) $\frac{6}{30} =$

19) $\frac{67}{1,000} =$

20) $\frac{1}{10} =$

21) $\frac{7}{20} =$

22) $\frac{4}{100} =$

Convert Decimal to Percent

Write each as a percent.

1) 0.165 =

2) 0.15 =

3) 1.4 =

4) 0.015 =

5) 0.005 =

6) 0.625 =

7) 0.185 =

8) 0.34 =

9) 0.03 =

10) 0.1 =

11) 0.175 =

12) 4.95 =

13) 2.105 =

14) 0.2 =

15) 1.05 =

16) 0.0275 =

17) 0.0015 =

18) 0.720 =

19) 2.25 =

20) 0.333 =

21) 6.175 =

22) 0.326 =

23) 1.8 =

24) 0.5 =

25) 1.5 =

26) 12.5 =

27) 3.05 =

28) 0.01 =

Convert Fraction to Percent

Write each as a percent.

1) $\dfrac{1}{5} =$

2) $\dfrac{5}{4} =$

3) $\dfrac{8}{16} =$

4) $\dfrac{19}{22} =$

5) $\dfrac{14}{20} =$

6) $\dfrac{13}{50} =$

7) $\dfrac{7}{9} =$

8) $\dfrac{13}{20} =$

9) $\dfrac{5}{100} =$

10) $\dfrac{8}{20} =$

11) $\dfrac{3}{25} =$

12) $\dfrac{14}{100} =$

13) $\dfrac{48}{50} =$

14) $\dfrac{32}{50} =$

15) $\dfrac{19}{28} =$

16) $\dfrac{3}{33} =$

17) $\dfrac{24}{44} =$

18) $\dfrac{23}{28} =$

19) $\dfrac{24}{84} =$

20) $\dfrac{5}{50} =$

21) $\dfrac{25}{625} =$

22) $\dfrac{480}{240} =$

Answer key Chapter 3

Round Decimals
1) 0.6
2) 2.0
3) 6.6
4) 0.4
5) 8.0
6) 0.09
7) 6.2
8) 76.76
9) 1.63
10) 10.4
11) 1.1
12) 4.2
13) 3.2
14) 6.05
15) 64.0
16) 37.0
17) 420.0
18) 512.7
19) 12.36
20) 66.0
21) 3.21
22) 37.1
23) 126.5
24) 0.01
25) 0.079
26) 5.02
27) 23.61
28) 8.08

Decimals Addition
1) 77.77
2) 1
3) 31.83
4) 59.304
5) 12.83
6) 7.379
7) 85.59
8) 79.71
9) 54.93
10) 20.831
11) 30.379
12) 74.35
13) 35.42
14) 128.64

Decimals Subtraction
1) 6.15
2) 16.45
3) 0.45
4) 2.584
5) 0.734
6) 60.81
7) 42.29
8) 15.98
9) 14.99
10) 5.575
11) 42.63
12) 13.418
13) 20.685
14) 111.79

Decimals Multiplication
1) 5.46
2) 51.33
3) 14.473
4) 670.8
5) 1,3080
6) 250.068
7) 313.7012
8) 28.623
9) 0.903
10) 191.2863
11) 224.51
12) 0.1536
13) 1.6905
14) 60.9636

Decimal Division
1) 0.007
2) 0.3
3) 0.0026

4) 0.0001
5) 0.142…
6) 0.024….
7) 0.0625
8) 0.0666…
9) 0.08
10) 1.111…
11) 10
12) 2.5
13) 1.4285…
14) 10
15) 0.207
16) 0.076
17) 0.00738
18) 3.4666…
19) 9.04
20) 10
21) 3.22
22) 0.25
23) 0.159…
24) 0.0016
25) 5.2
26) 0.0002

Comparing Decimals
1) <
2) >
3) =
4) <
5) >
6) <
7) >
8) <
9) >
10) >
11) <
12) <
13) <
14) >
15) >
16) =
17) <
18) <
19) <
20) >
21) <
22) >
23) >
24) >
25) >
26) =
27) >
28) <

Convert Fraction to Decimal
1) 0.4
2) 0.38
3) 0.16
4) 0.25
5) 0.11
6) 0.49
7) 0.08
8) 0.68
9) 0.235
10) 0.26
11) 0.5
12) 0.375
13) 0.3
14) 0.072
15) 0.54
16) 0.4
17) 4.5
18) 0.2
19) 0.067
20) 0.1
21) 0.35
22) 0.04

Convert Decimal to Percent
1) 16.5%
2) 15%
3) 140%
4) 1.5%
5) 0.5%
6) 62.5%
7) 18.5%
8) 34%
9) 3%

10) 10%
11) 17.5%
12) 495%
13) 210.5%
14) 20%
15) 105%
16) 2.75%

17) 0.15%
18) 72%
19) 225%
20) 33.3%
21) 617.5%
22) 32.6%
23) 180%

24) 50%
25) 150%
26) 1,250%
27) 305%
28) 1%

Convert Fraction to Percent

1) 20%
2) 125%
3) 50%
4) 86.36%
5) 70%
6) 26%
7) 77.8%
8) 65%

9) 5%
10) 40%
11) 12%
12) 14%
13) 96%
14) 64%
15) 67.9%
16) 9.09%

17) 54.5%
18) 82.14%
19) 28.57%
20) 10%
21) 4%
22) 200%

Chapter 4:
Exponent and Radicals

Positive Exponents

Simplify. Your answer should contain only positive exponents.

1) $2^3 =$

2) $5^3 =$

3) $\frac{2x^5y}{xy} =$

4) $(15x3x)^2 =$

5) $(x^3)^2 =$

6) $(\frac{1}{5})^2 =$

7) $0^6 =$

8) $5 \times 5 \times 5 =$

9) $2 \times 2 \times 2 \times 2 \times 2 =$

10) $(3x^2y)^3 =$

11) $10^3 =$

12) $(2x^2y^4)^3 =$

13) $4 \times 10^3 =$

14) $0.5 \times 0.5 \times 0.5 =$

15) $\frac{1}{2} \times \frac{1}{2} \times \frac{1}{2} =$

16) $3^3 =$

17) $(10x^{10}y^3)^2 =$

18) $2^5 =$

19) $x \times x \times x =$

20) $3 \times 3 \times 3 \times 3 \times 3 =$

21) $(3x^2y^3z)^2 =$

22) $7^0 =$

23) $(12x^5y^{-2})^2 =$

24) $(3x^3y^2)^4 =$

WWW.MathNotion.com

Negative Exponents

Simplify. Leave no negative exponents.

1) $3^{-2} =$

2) $7^{-1} =$

3) $\left(\frac{1}{5}\right)^{-3} =$

4) $10^{-5} =$

5) $1^{-100} =$

6) $4^{-4} =$

7) $\left(\frac{1}{2}\right)^{-3} =$

8) $-5y^{-3} =$

9) $\left(\frac{1}{y^{-4}}\right)^{-2} =$

10) $x^{-\frac{3}{2}} =$

11) $\frac{1}{2^{-5}} =$

12) $3^{-4} =$

13) $2^{-3} =$

14) $15^{-1} =$

15) $20^{-2} =$

16) $x^{-4} =$

17) $(x^3)^{-2} =$

18) $x^{-1} \times x^{-1} \times x^{-1} =$

19) $\frac{1}{2} \times \frac{1}{2} =$

20) $10^{-2} =$

21) $10z^{-2} =$

22) $2^{-5} =$

23) $\left(-\frac{1}{3}\right)^4 =$

24) $6^0 =$

25) $\left(\frac{1}{x}\right)^{-4} =$

26) $12^{-2} =$

Add and subtract Exponents

Solve each problem.

1) $3^2 + 2^5 =$

2) $x^6 + x^6 =$

3) $3b^2 - 2b^2 =$

4) $3 + 4^3 =$

5) $8 - 4^2 =$

6) $4 + 7^1 =$

7) $2x^3 + 3x^3 =$

8) $10^2 + 3^5 =$

9) $4^5 - 2^4 =$

10) $5^2 - 6^0 =$

11) $1^2 - 3^0 =$

12) $7^1 + 2^3 =$

13) $6^1 - 5^3 =$

14) $3^3 + 3^3 =$

15) $9^2 - 8^2 =$

16) $0^{73} + 0^{54} =$

17) $2^2 - 3^2 =$

18) $7^3 - 7^1 =$

19) $8^2 - 6^2 =$

20) $4^2 + 3^2 =$

21) $2^3 + 4^3 =$

22) $10 + 3^3 =$

23) $6x^5 + 8x^5 =$

24) $8^0 + 4^2 =$

25) $3^2 + 3^2 =$

26) $10^2 + 5^2 =$

27) $(\frac{1}{2})^2 + (\frac{1}{2})^2 =$

28) $9^2 + 3^2 =$

Exponent multiplication

Simplify each of the following

1) $3^6 \times 3^2 =$

2) $9^2 \times 5^0 =$

3) $6^1 \times 7^3 =$

4) $a^{-3} \times a^{-3} =$

5) $y^{-2} \times y^{-2} \times y^{-2} =$

6) $2^4 \times 3^4 \times 2^{-2} \times 3^{-3} =$

7) $5x^2y^3 \times 8x^3y^5 =$

8) $(x^2)^3 =$

9) $(x^2y^3)^4 \times (x^2y^4)^{-4} =$

10) $6^3 \times 6^2 =$

11) $a^{2b} \times a^0 =$

12) $2^3 \times 2^4 =$

13) $a^m \times a^n =$

14) $a^n \times b^n =$

15) $6^{-2} \times 3^{-2} =$

16) $5^{12} \times 2^{12} =$

17) $(3^5)^4 =$

18) $(\frac{1}{5})^3 \times (\frac{1}{5})^2 \times (\frac{1}{5})^4 =$

19) $(\frac{1}{7})^{32} \times 7^{32} =$

20) $(2m)^{\frac{2}{3}} \times (-3m)^{\frac{2}{3}} =$

21) $(x^2y^3)^{\frac{1}{5}} \times (x^2y^2)^{\frac{1}{5}} =$

22) $(a^m b^n)^r =$

23) $(3x^2y^3)^4 =$

24) $(x^{\frac{1}{2}}y^3)^{\frac{-1}{2}} \times (x^2y^4)^0 =$

25) $6^3 \times 6^4 =$

26) $32^{\frac{1}{4}} \times 32^{\frac{1}{2}} =$

27) $8^4 \times 2^4 =$

28) $(x^3)^0 =$

Exponent division

Simplify. Your answer should contain only positive exponents.

1) $\dfrac{4^3}{4} =$

2) $\dfrac{25x^3}{x} =$

3) $\dfrac{a^m}{a^n} =$

4) $\dfrac{2x^{-5}}{10x^{-3}} =$

5) $\dfrac{81x^8}{9x^3} =$

6) $\dfrac{11x^6}{4x^7} =$

7) $\dfrac{18x^2}{6y^5} =$

8) $\dfrac{35xy^5}{x^5y^2} =$

9) $\dfrac{2x^5}{7x} =$

10) $\dfrac{36x^3y^7}{4x^4} =$

11) $\dfrac{9x^2}{15x^7y^9} =$

12) $\dfrac{yx^4}{5yx^7} =$

13) $\dfrac{14x^2y}{2xy^2} =$

14) $\dfrac{x^{3.25}}{x^{0.25}} =$

15) $\dfrac{5x^3y}{10xy^2} =$

16) $\dfrac{16ab^2r^9}{8a^3b^4} =$

17) $\dfrac{20x^3}{10x^5} =$

18) $\dfrac{16x^3}{4x^6} =$

19) $\dfrac{5^4}{5^2} =$

20) $\dfrac{x}{x^{12}} =$

21) $\dfrac{10^6}{10^2} =$

22) $\dfrac{2xy^4}{8y^2} =$

23) $\dfrac{12x^5y}{144xy^2} =$

24) $\dfrac{42x^6}{7y^8} =$

Scientific Notation

Write each number in scientific notation.

1) $8,100,000 =$

2) $50 =$

3) $0.0000008 =$

4) $254,000 =$

5) $0.000225 =$

6) $6.5 =$

7) $0.00063 =$

8) $19,000,000 =$

9) $5,000,000 =$

10) $85,000,000 =$

11) $0.0000036 =$

12) $0.00012 =$

13) $0.005 =$

14) $6,600 =$

15) $1,960 =$

16) $170,000 =$

17) $0.115 =$

18) $0.05 =$

19) $0.0033 =$

20) $20,000 =$

21) $23,000 =$

22) $0.00000102 =$

23) $0.0102 =$

24) $1,568 =$

25) $32,581 =$

26) $12,500 =$

27) $12,054 =$

28) $60,000 =$

Square Roots

Find the square root of each number.

1) $\sqrt{1} =$

2) $\sqrt{4} =$

3) $\sqrt{16} =$

4) $\sqrt{25} =$

5) $\sqrt{49} =$

6) $\sqrt{81} =$

7) $\sqrt{100} =$

8) $\sqrt{144} =$

9) $\sqrt{121} =$

10) $\sqrt{169} =$

11) $\sqrt{9} =$

12) $\sqrt{36} =$

13) $\sqrt{225} =$

14) $\sqrt{196} =$

15) $\sqrt{256} =$

16) $\sqrt{625} =$

17) $\sqrt{289} =$

18) $\sqrt{1,024} =$

19) $\sqrt{484} =$

20) $\sqrt{361} =$

21) $\sqrt{441} =$

22) $\sqrt{841} =$

23) $\sqrt{729} =$

24) $\sqrt{900} =$

25) $\sqrt{400} =$

26) $\sqrt{3,600} =$

27) $\sqrt{4,900} =$

28) $\sqrt{6,400} =$

Simplify Square Roots

Simplify the following.

1) $\sqrt{72} =$

2) $\sqrt{27} =$

3) $\sqrt{28} =$

4) $\sqrt{44} =$

5) $\sqrt{50} =$

6) $\sqrt{40} =$

7) $10\sqrt{125} =$

8) $5\sqrt{600} =$

9) $\sqrt{18} =$

10) $3\sqrt{32} =$

11) $2\sqrt{5} + 8\sqrt{5} =$

12) $\frac{1}{1+\sqrt{2}} =$

13) $\sqrt{20} =$

14) $\frac{5}{2-\sqrt{3}} =$

15) $\sqrt{3} \times \sqrt{12} =$

16) $\frac{\sqrt{400}}{\sqrt{4}} =$

17) $\frac{\sqrt{48}}{\sqrt{16 \times 3}} =$

18) $\sqrt{24y^4} =$

19) $7\sqrt{64a} =$

20) $\sqrt{4+32} + \sqrt{16} =$

21) $\sqrt{90} =$

22) $\sqrt{338} =$

23) $\sqrt{60} =$

24) $\sqrt{75} =$

25) $\sqrt{1,875} =$

26) $\sqrt{32} =$

Answer key Chapter 4

Positive Exponents

1) 8
2) 125
3) $2x^4$
4) $2,025x^4$
5) x^6
6) $\frac{1}{25}$
7) 0
8) 5^3
9) 2^5
10) $27x^6y^3$
11) 1,000
12) $8x^6y^{12}$
13) 4,000
14) 0.5^3
15) $(\frac{1}{2})^3$
16) 27
17) $100x^{20}y^6$
18) 32
19) x^3
20) 3^5
21) $9x^4y^6z^2$
22) 1
23) $\frac{144x^{10}}{y^4}$
24) $81x^{12}y^8$

Negative Exponents

1) $\frac{1}{9}$
2) $\frac{1}{7}$
3) 125
4) $\frac{1}{100,000}$
5) 1
6) $\frac{1}{256}$
7) 8
8) $\frac{-5}{y^3}$
9) y^8
10) $\frac{1}{x^{\frac{3}{2}}}$
11) 2^5
12) $\frac{1}{81}$
13) $\frac{1}{8}$
14) $\frac{1}{15}$
15) $\frac{1}{400}$
16) $\frac{1}{x^4}$
17) $\frac{1}{x^6}$
18) $\frac{1}{x^3}$
19) $\frac{1}{2^2}$
20) $\frac{1}{100}$
21) $\frac{10}{z^2}$
22) $\frac{1}{32}$
23) $\frac{1}{81}$
24) 1
25) x^4
26) $\frac{1}{144}$

Add and subtract Exponents

1) 41
2) $2x^6$
3) b^2
4) 67
5) −8
6) 11
7) $5x^3$
8) 343
9) 1,008
10) 24
11) 0
12) 15
13) −119
14) 54
15) 17
16) 0
17) −5
18) 336

Common Core Math Practice Grade 7

19) 28
20) 25
21) 72
22) 37
23) $14x^5$
24) 17
25) 18
26) 125
27) $\frac{1}{2}$
28) 90

Exponent multiplication

1) 3^8
2) 81
3) 2,058
4) a^{-6}
5) y^{-6}
6) $2^2 \times 3^1 = 12$
7) $40x^5y^8$
8) x^6
9) y^{-4}
10) 6^5
11) a^{2b}
12) 2^7
13) a^{m+n}
14) $(ab)^n$
15) 18^{-2}
16) 10^{12}
17) 3^{20}
18) $(\frac{1}{5})^9$
19) 1
20) $(-6m)^{\frac{2}{3}}$
21) $x^{\frac{4}{5}}y$
22) $a^{mr}b^{nr}$
23) $81x^8y^{12}$
24) $x^{\frac{-1}{4}}y^{\frac{-3}{2}}$
25) 6^7
26) $32^{\frac{3}{4}}$
27) $16^4 = 2^{16}$
28) 1

Exponent division

1) 4^2
2) $25x^2$
3) a^{m-n}
4) $\frac{1}{5x^2}$
5) $9x^5$
6) $\frac{11}{4x}$
7) $\frac{3x^2}{y^5}$
8) $\frac{35y^3}{x^4}$
9) $\frac{2x^4}{7}$
10) $\frac{9y^7}{x}$
11) $\frac{3}{5x^5y^9}$
12) $\frac{1}{5x^3}$
13) $\frac{7x}{y}$
14) x^3
15) $\frac{x^2}{2y}$
16) $\frac{2r^9}{a^2b^2}$
17) $\frac{2}{x^2}$
18) $\frac{4}{x^3}$
19) 5^2
20) $\frac{1}{x^{11}}$
21) 10^4
22) $\frac{1}{4}xy^2$
23) $\frac{x^4}{12y}$
24) $\frac{6x^6}{y^8}$

Scientific Notation

1) 81×10^5
2) 5×10^1
3) 8×10^{-7}
4) 2.54×10^5
5) 2.25×10^{-4}
6) 65×10^{-1}
7) 63×10^{-5}
8) 1.9×10^7
9) 5×10^6

Common Core Math Practice Grade 7

10) 8.5×10^7
11) 3.6×10^{-6}
12) 1.2×10^{-4}
13) 5×10^{-3}
14) 6.6×10^3
15) 1.96×10^3
16) 1.7×10^5

17) 1.15×10^{-1}
18) 5×10^{-2}
19) 33×10^{-4}
20) 2×10^4
21) 23×10^3
22) 102×10^{-8}
23) 1.02×10^{-2}

24) 1.568×10^3
25) 32.581×10^3
26) 12.5×10^3
27) 1.2054×10^4
28) 6×10^4

Square Roots

1) 1
2) 2
3) 4
4) 5
5) 7
6) 9
7) 10
8) 12
9) 11
10) 13

11) 3
12) 6
13) 15
14) 14
15) 16
16) 25
17) 17
18) 32
19) 22
20) 19

21) 21
22) 29
23) 27
24) 30
25) 20
26) 60
27) 70
28) 80

Simplify Square Roots

1) $6\sqrt{2}$
2) $3\sqrt{3}$
3) $2\sqrt{7}$
4) $2\sqrt{11}$
5) $5\sqrt{2}$
6) $2\sqrt{10}$
7) $50\sqrt{5}$
8) $50\sqrt{6}$
9) $3\sqrt{2}$

10) $12\sqrt{2}$
11) $10\sqrt{5}$
12) $\sqrt{2} - 1$
13) $2\sqrt{5}$
14) $10 + 5\sqrt{3}$
15) 6
16) 10
17) 1
18) $2y^2\sqrt{6}$

19) $56\sqrt{a}$
20) 10
21) $3\sqrt{10}$
22) $13\sqrt{2}$
23) $2\sqrt{15}$
24) $5\sqrt{3}$
25) $25\sqrt{3}$
26) $4\sqrt{2}$

Chapter 5: Ratio, Proportion and Percent

Proportions

Find a missing number in a proportion.

1) $\dfrac{5}{8} = \dfrac{20}{a}$

2) $\dfrac{a}{6} = \dfrac{24}{36}$

3) $\dfrac{14}{42} = \dfrac{a}{3}$

4) $\dfrac{15}{a} = \dfrac{75}{32}$

5) $\dfrac{8}{a} = \dfrac{32}{150}$

6) $\dfrac{\sqrt{16}}{5} = \dfrac{a}{30}$

7) $\dfrac{5}{12} = \dfrac{15}{a}$

8) $\dfrac{6}{12} = \dfrac{a}{33.6}$

9) $\dfrac{8}{a} = \dfrac{3.2}{4}$

10) $\dfrac{1}{16} = \dfrac{3}{a}$

11) $\dfrac{10}{8} = \dfrac{5}{a}$

12) $\dfrac{12}{a} = \dfrac{3}{17}$

13) $\dfrac{2}{7} = \dfrac{a}{10}$

14) $\dfrac{\sqrt{25}}{4} = \dfrac{30}{a}$

15) $\dfrac{12}{a} = \dfrac{13.2}{19.8}$

16) $\dfrac{50}{190} = \dfrac{a}{380}$

17) $\dfrac{32}{100} = \dfrac{a}{52}$

18) $\dfrac{27}{81} = \dfrac{a}{3}$

19) $\dfrac{5}{8} = \dfrac{1}{a}$

20) $\dfrac{5}{3} = \dfrac{35}{a}$

Reduce Ratio

Reduce each ratio to the simplest form.

1) 3: 12 =

2) 4: 24 =

3) 81: 45 =

4) 30: 25 =

5) 24: 240 =

6) 80: 10 =

7) 80: 400 =

8) 5: 180 =

9) 24: 72 =

10) 3.6: 4.2 =

11) 220: 660 =

12) 1.8: 3 =

13) 150: 250 =

14) 40: 60 =

15) 26: 52 =

16) 16: 4 =

17) 100: 25 =

18) 10: 100 =

19) 108: 72 =

20) 130: 165 =

21) 30: 60 =

22) 24: 28 =

23) 10: 150 =

24) 15: 90 =

Percent

Find the Percent of Numbers.

1) 20% of 38 =

2) 42% of 7 =

3) 11% of 11 =

4) 36% of 75 =

5) 5% of 50 =

6) 32% of 14 =

7) 12% of 3 =

8) 9% of 47 =

9) 50% of 52 =

10) 7.5% of 60 =

11) 92% of 12 =

12) 80% of 60 =

13) 12% of 120 =

14) 1% of 310 =

15) 32% of 0 =

16) 62% of 100 =

17) 32% of 44 =

18) 15% of 60 =

19) 5% of 10 =

20) 3% of 7 =

21) 40% of 20 =

22) 70% of 2 =

23) 25% of 20 =

24) 7% of 200 =

25) 50% of 300 =

26) 3% of 6 =

27) 6% of 400 =

28) 9% of 6 =

Discount, Tax and Tip

Find the selling price of each item.

1) Original price of a computer: $250
   Tax: 6%, Selling price: $_____

2) Original price of a laptop: $320
   Tax: 5%, Selling price: $_____

3) Original price of a sofa: $400
   Tax: 7%, Selling price: $_____

4) Original price of a car: $16,500
   Tax: 4.5%, Selling price: $_____

5) Original price of a Table: $300
   Tax: 6%, Selling price: $_____

6) Original price of a house: $450,000
   Tax: 2.5%, Selling price: $_____

7) Original price of a tablet: $200
   Discount: 20%, Selling price: $_____

8) Original price of a chair: $250
   Discount: 15%, Selling price: $_____

9) Original price of a book: $50
   Discount: 35% Selling price: $_____

10) Original price of a cellphone: 600
    Discount: 10% Selling price: $_____

11) Food bill: $32
    Tip: 20%   Price: $_____

12) Food bill: $30
    Tipp: 15%   Price: $_____

13) Food bill: $64
    Tip: 20%   Price: $_____

14) Food bill: $36
    Tipp: 25%   Price: $_____

Find the answer for each word problem.

15) Nicolas hired a moving company. The company charged $200 for its services, and Nicolas gives the movers a 30% tip. How much does Nicolas tip the movers? $_____

16) Mason has lunch at a restaurant and the cost of his meal is $60. Mason wants to leave a 10% tip. What is Mason's total bill including tip? $_____

Percent of Change

Find each percent of change.

1) From 200 to 400. ___%

2) From 25 ft to 125 ft. ___%

3) From $50 to $350. ___%

4) From 40 cm to 160 cm. ___%

5) From 20 to 60. ___%

6) From 40 to 8. ___%

7) From 160 to 240. ___%

8) From 600 to 300. ___%

9) From 75 to 45. ___%

10) From 128 to 32. ___%

Calculate each percent of change word problem.

11) Bob got a raise, and his hourly wage increased from $24 to $30. What is the percent increase? ____%

12) The price of a pair of shoes increases from $60 to $96. What is the percent increase? ___%

13) At a coffeeshop, the price of a cup of coffee increased from $2.40 to $2.88. What is the percent increase in the cost of the coffee? _____%

14) 24cm are cut from a 96 cm board. What is the percent decrease in length? _ %

15) In a class, the number of students has been increased from 108 to 162. What is the percent increase? _____%

16) The price of gasoline rose from $16.80 to $19.32 in one month. By what percent did the gas price rise? _____%

17) A shirt was originally priced at $24. It went on sale for $19.20. What was the percent that the shirt was discounted? _____%

Simple Interest

Determine the simple interest for these loans.

1) $225 at 14% for 2 years. $ _____
2) $2,600 at 8% for 3 years. $ _____
3) $1,300 at 15% for 5 years. $ _____
4) $8,400 at 2.5% for 5 months. $ ___
5) $300 at 2% for 9 months. $ _____

6) $48,000 at 5.5% for 5 years. $ ____
7) $5,200 at 9% for 2 years. $ _____
8) $600 at 5.5% for 4 years. $ _____
9) $800 at 4.5 % for 9 months. $ ____
10) $6,000 at 2.2% for 5 years. $ ___

Calculate each simple interest word problem.

11) A new car, valued at $14,000, depreciates at 4.5% per year. What is the value of the car one year after purchase? $_____

12) Sara puts $8,000 into an investment yielding 5% annual simple interest; she left the money in for two years. How much interest does Sara get at the end of those two years? $_____

13) A bank is offering 10.5% simple interest on a savings account. If you deposit $22,500, how much interest will you earn in two years? $_____

14) $800 interest is earned on a principal of $8,000 at a simple interest rate of 5% interest per year. For how many years was the principal invested? _____

15) In how many years will $1,500 yield an interest of $300 at 5% simple interest? _____

16) Jim invested $6,000 in a bond at a yearly rate of 3.5%. He earned $630 in interest. How long was the money invested? _____

WWW.MathNotion.com

Answer key Chapter 5

Proportions

1) $a = 32$
2) $a = 4$
3) $a = 1$
4) $a = 6.4$
5) $a = 37.5$
6) $a = 24$
7) $a = 36$
8) $a = 16.8$
9) $a = 10$
10) $a = 48$
11) $a = 4$
12) $a = 68$
13) $a = \frac{20}{7}$
14) $a = 24$
15) $a = 18$
16) $a = 100$
17) $a = 16.64$
18) $a = 1$
19) $a = 1.6$
20) $a = 21$

Reduce Ratio

1) 1: 4
2) 1: 6
3) 9: 5
4) 6: 5
5) 1: 10
6) 8: 1
7) 1: 5
8) 1: 36
9) 1: 3
10) 0.6: 0.7
11) 11: 33
12) 0.6: 1
13) 3: 5
14) 2: 3
15) 1: 2
16) 4: 1
17) 4: 1
18) 1: 10
19) 3: 2
20) 26: 33
21) 1: 2
22) 6: 7
23) 1: 15
24) 1: 6

Percent

1) 7.6
2) 2.94
3) 1.21
4) 27
5) 2.5
6) 4.48
7) 0.36
8) 4.23
9) 26
10) 4.5
11) 11.04
12) 48
13) 14.4
14) 3.1
15) 0
16) 62
17) 14.08
18) 9
19) 0.5
20) 0.21
21) 8
22) 1.4
23) 5
24) 14
25) 150
26) 0.18
27) 24
28) 0.54

Discount, Tax and Tip

1) $265.00
2) $336.00
3) $428.00
4) $17,242.50
5) $318.00
6) $461,250
7) $240.00
8) $287.50
9) $67.50
10) $660.00
11) $38.40
12) $34.50
13) $76.80
14) $45.00
15) $60.00
16) $66.00

Percent of Change

1) 100%
2) 400%
3) 600%
4) 300%
5) 200%
6) 80%
7) 50%
8) 50%
9) 40%
10) 75%
11) 25%
12) 60%
13) 20%
14) 25%
15) 50%
16) 15%
17) 20%

Simple Interest

1) $63.00
2) $624.00
3) $975.00
4) $87.50
5) $4.50
6) $13,200.00
7) $936.00
8) $132.00
9) $27.00
10) $660.00
11) $13,370.00
12) $800.00
13) $4725.00
14) 2 *years*
15) 4 *years*
16) 3 *years*

Chapter 6:

Measurement

Reference Measurement

LENGTH	
Customary	**Metric**
1 mile (mi) = 1,760 yards (yd)	1 kilometer (km) = 1,000 meters (m)
1 yard (yd) = 3 feet (ft)	1 meter (m) = 100 centimeters (cm)
1 foot (ft) = 12 inches (in.)	1 centimeter(cm) = 10 millimeters(mm)
VOLUME AND CAPACITY	
Customary	**Metric**
1 gallon (gal) = 4 quarts (qt)	1 liter (L) = 1,000 milliliters (mL)
1 quart (qt) = 2 pints (pt.)	
1 pint (pt.) = 2 cups (c)	
1 cup (c) = 8 fluid ounces (Fl oz)	
WEIGHT AND MASS	
Customary	**Metric**
1 ton (T) = 2,000 pounds (lb.)	1 kilogram (kg) = 1,000 grams (g)
1 pound (lb.) = 16 ounces (oz)	1 gram (g) = 1,000 milligrams (mg)
Time	
1 year = 12 months	
1 year = 52 weeks	
1 week = 7 days	
1 day = 24 hours	
1 hour = 60 minutes	
1 minute = 60 seconds	

Metric Length Measurement

Convert to the units.

1) 5×10^4 mm = _____ cm

2) 0.4 m = _____ mm

3) 0.06 m = _____ cm

4) 1.2 km = _____ m

5) 8,000 mm = _____ m

6) 4,700 cm = _____ m

7) 4.5 m = _____ cm

8) 7×10^3 mm = _____ cm

9) 9×10^6 mm = _____ m

10) 2 km = _____ mm

11) 0.3 km = _____ m

12) 0.05 m = _____ cm

13) 4×10^4 m = _____ km

14) 6×10^7 m = _____ km

Customary Length Measurement

Convert to the units.

1) 20 ft = _____ in

2) 2.5 ft = _____ in

3) 5.6 yd = _____ ft

4) 0.4 yd = _____ ft

5) 9×10^{-1} yd = _____ in

6) 2 mi = _____ in

7) 18×10^3 in = _____ yd

8) 21.6 in = _____ yd

9) 6,160 yd = _____ mi

10) 28 yd = _____ in

11) 0.03 mi = _____ yd

12) 99×10^3 ft = _____ mi

13) 4.8 in = _____ ft

14) 42 yd = _____ feet

15) 0.72 in = _____ ft

16) 0.2 mi = _____ ft

Metric Capacity Measurement

Convert the following measurements.

1) 60 l = _____ ml

2) 0.7 l = _____ ml

3) 2.8 l = _____ ml

4) 0.06 l = _____ ml

5) 22.5 l = _____ ml

6) 0.9 l = _____ ml

7) 6×10^6 ml = _____ l

8) 22×10^5 ml = _____ l

9) 112×10^2 ml = _____ l

10) 11,000 ml = _____ l

11) 57,800 ml = _____ l

12) 0.3×10^5 ml = _____ l

Customary Capacity Measurement

Convert the following measurements.

1) 1.5 gal = _____ qt.

2) 4.5 gal = _____ pt.

3) 0.5 gal = _____ c.

4) 18 pt. = _____ c

5) 12 c = _____ fl oz

6) 8.15 qt = _____ pt.

7) 0.08 qt = _____ c

8) 42 pt. = _____ c

9) 8×10^4 c = _____ gal

10) 256 pt. = _____ gal

11) 484 qt = _____ gal

12) 25.8 pt. = _____ qt

13) 7×10^3 c = _____ qt

14) 98.8 c = _____ pt.

15) 0.164 qt = _____ gal

16) 1,256 pt. = _____ qt

17) 23 gal = _____ pt.

18) 0.01 qt = _____ c

19) 800 c = _____ gal

20) 64.16 fl oz = _____ c

Metric Weight and Mass Measurement

Convert.

1) 0.8 kg = _____ g

2) 5.6 kg = _____ g

3) 2×10^{-4} kg = _____ g

4) 1.04 kg = _____ g

5) 44.8 kg = _____ g

6) 13.12 kg = _____ g

7) 0.072 kg = _____ g

8) 21×10^5 g = _____ kg

9) 15×10^6 g = _____ kg

10) 0.04×10^8 g = _____ kg

11) 17,400 g = _____ kg

12) 98×10^2 g = _____ kg

13) 5,400,000 g = _____ kg

14) 325×10^4 g = _____ kg

Customary Weight and Mass Measurement

Convert.

1) 24×10^4 lb. = _____ T

2) 0.32×10^5 lb. = _____ T

3) 190,000 lb. = _____ T

4) 2,800 lb. = _____ T

5) 0.35 lb. = _____ oz

6) 2.8 lb. = _____ oz

7) 0.05 lb. = _____ oz

8) 4 T = _____ lb.

9) 7×10^{-4} T = _____ lb.

10) 38×10^{-5} T = _____ lb.

11) 0.6 T = _____ lb.

12) 0.003 T = _____ oz

13) 0.015 T = _____ oz

14) 196.8 oz = _____ lb.

Unit of Measurements

Use the given ratios to convert the measuring units. If necessary, round the answers to three decimal digits.

1) Use $1 = \frac{1.6093 \text{km}}{1 \text{ mi}}$ and convert 6.02 miles to kilometers

   6.02 mi = _____

2) Use $1 = \frac{1.6093 \text{km}}{1 \text{ mi}}$ and convert 4.15 miles to kilometers

   4.15 mi = _____

3) Use $1 = \frac{1 \text{qt}}{0.946 \text{L}}$ and convert 6 liters to quarts

   6 L = _____

4) Use $1 = \frac{1 \text{qt}}{0.946 \text{L}}$ and convert 8 liters to quarts

   8 L = _____

5) Use $1 = \frac{1.6093 \text{km}}{1 \text{ mi}}$ and convert 5.06 miles to kilometers

   5.06 mi = _____

6) Use $1 = \frac{1.6093 \text{km}}{1 \text{ mi}}$ and convert 8.1 miles to kilometers

   8.1 mi = _____

7) Use $1 = \frac{1 \text{qt}}{0.946 \text{L}}$ and convert 5 liters to quarts

   5 L = _____

Temperature

Convert Fahrenheit into Celsius.

1) 176°F = ___ °C

2) 113°F= ___ °C

3) 122°F= ___ °C

4) −13°F= ___ °C

5) 131°F= ___ °C

6) 136.4°F= ___ °C

7) 14°F= ___ °C

8) 149°F= ___ °C

9) 158°F= ___ °C

10) 167°F= ___ °C

11) 77°F= ___ °C

12) 86°F= ___ °C

Convert Celsius into Fahrenheit.

13) 85°C= ___ °F

14) 150°C= ___ °F

15) 83°C= ___ °F

16) 20°C= ___ °F

17) 5°C= ___ °F

18) −5°C= ___ °F

19) 0°C= ___ °F

20) 30°C= ___ °F

21) 90°C= ___ °F

22) 72°C= ___ °F

23) 38°C= ___ °F

24) 35°C= ___ °F

Time

Convert to the units.

1) 28 hr. = _____ min

2) 15 year = _____ week

3) 0.5 hr. = _____ sec

4) 8.5 min = _____ sec

5) 6×10^4 min = _____ hr

6) 1,095 day = _____ year

7) 2 year = _____ hr.

8) 42 day = _____ hr

9) 2 day = _____ min

10) 480 min = _____ hr

11) 28.5 year = _____ month

12) 12,600 sec = _____ min

13) 216 hr = _____ day

14) 15 weeks = _____ day

How much time has passed?

1) From 3:35 A.M. to 6:45 A.M.: ____ hours and ____ minutes.

2) From 2:30 A.M. to 7:15 A.M.: ____ hours and ____ minutes.

3) It's 6:20 P.M. What time was 3 hours ago? _____ O'clock

4) 4:15 A.M to 7:35 AM: _____ hours and _____ minutes.

5) 1:45 A.M to 4:20 AM: _____ hours and _____ minutes.

6) 9:00 A.M. to 10:05 AM. = _____ hour(s) and _____ minutes.

7) 10:35 A.M. to 3:05 PM. = _____ hour(s) and _____ minutes

8) 5:12 A.M. to 5:48 A.M. = _____ minutes

9) 8:08 A.M. to 8:45 A.M. = _____ minutes

Answers of Worksheets – Chapter 6

Metric length

1) 5,000 cm
2) 400 mm
3) 6 cm
4) 1,200 m
5) 8 m
6) 47 m
7) 450 cm
8) 700 cm
9) 9,000 m
10) 2,000,000 mm
11) 300 m
12) 5 cm
13) 40 km
14) 60,000 km

Customary Length

1) 240
2) 30
3) 16.8
4) 1.2
5) 32.4
6) 126,720
7) 500
8) 0.6
9) 3.5
10) 1,008
11) 52.8
12) 18.75
13) 0.4
14) 126
15) 0.06
16) 1,056

Metric Capacity

1) 60,000 ml
2) 700 ml
3) 2,800 ml
4) 60 ml
5) 22,500 ml
6) 900 ml
7) 6,000 ml
8) 2,200 ml
9) 11.2 ml
10) 11 L
11) 57.8 L
12) 30 L

Customary Capacity

1) 6 qt
2) 36 pt.
3) 8 c
4) 36 c
5) 96 fl oz
6) 16.3 pt.
7) 0.32 c
8) 84 c
9) 5,000 gal
10) 32 gal
11) 121 gal
12) 12.9 qt
13) 1,750 qt
14) 49.4 pt.
15) 0.041 gal
16) 628 qt
17) 184 pt.
18) 0.04 c
19) 50 gal
20) 8.02 c

Metric Weight and Mass

1) 800 g
2) 5,600 g
3) 0.2 g
4) 1,040 g
5) 44,800 g
6) 13,120 g
7) 72 g
8) 2,100 kg
9) 15,000 kg
10) 4,000 kg
11) 17.4 kg
12) 9.8 kg
13) 5,400 kg
14) 3,250 kg

Common Core Math Practice Grade 7

Customary Weight and Mass

1) 120 T
2) 16 T
3) 95 T
4) 1.4 T
5) 5.6 oz
6) 44.8 oz
7) 0.8 oz
8) 8,000 lb.
9) 1.4 lb.
10) 0.76 lb.
11) 1,200 lb.
12) 96 oz
13) 480 oz
14) 12.3 lb

Unit of measurements

1) 9.688km
2) 6.679km
3) 6.342 qt
4) 8.457qt
5) 8.143km
6) 13.035km
7) 5.285qt

Temperature

1) 80°C
2) 45°C
3) 50°C
4) −25°C
5) 55°C
6) 58°C
7) −10°C
8) 65°C
9) 70°C
10) 75°C
11) 25°C
12) 30°C
13) 185°F
14) 302°F
15) 181.4°F
16) 68°F
17) 41°F
18) 23°F
19) 32°F
20) 86°F
21) 194°F
22) 161.6°F
23) 100.4°F
24) 95°F

Time - Convert

1) 1,680 min
2) 780 weeks
3) 1,800 sec
4) 510 sec
5) 1,000 hr
6) 3 year
7) 17,520 hr
8) 1,008 hr
9) 2,880 min
10) 8 hr
11) 342 months
12) 210 min
13) 9 days
14) 105 days

Time - Gap

1) 3:10
2) 4:45
3) 3:20 P.M.
4) 3:20
5) 2:35
6) 1:05
7) 4:30
8) 36 minutes
9) 37 minutes

Chapter 7:
Linear Functions

Common Core Math Practice Grade 7

Relation and Functions

Determine whether each relation is a function. Then state the domain and range of each relation.

1)
Function: ....
Domain: ....
Range: ....

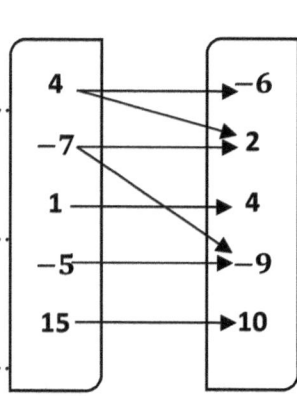

2)
Function: ....
Domain: ....
Range: ....

x	y
4	5
2	3
−6	−8
6	−8
−11	2

3)
Function: ....
Domain: ....
Range: ....

4) $\{(2,-2),(7,-6),(9,9),(8,1),(7,4)\}$
Function: ....
Domain: ....
Range: ....

5)
Function: ....
Domain: ....
Range: ....

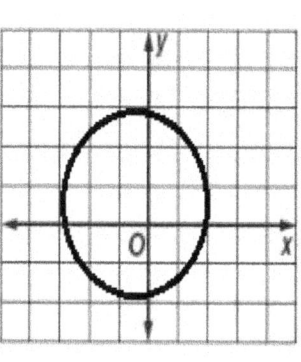

6)
Function: ....
Domain: ....
Range: ....

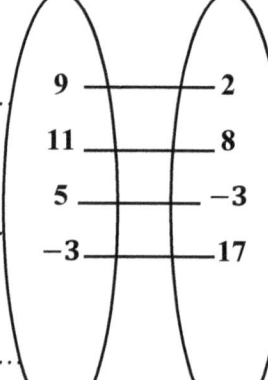

WWW.MathNotion.com

Common Core Math Practice Grade 7

Slope form

Write the slope-intercept form of the equation of each line.

1) $4x + 5y = 15$

2) $4x + 12y = 3$

3) $7x + y = -9$

4) $-7x + 11y = 5$

5) $5x - 4y = 7$

6) $-21x + 3y = 6$

7) $2x + y = 0$

8) $5x - 7y = -9$

9) $-13.5x + 27y = 54$

10) $-3x + \frac{2}{3}y = 18$

11) $10x + y = -120$

12) $3x = -36y - 27$

13) $1.5x = 3y + 3$

14) $5x = -\frac{5}{4}y + 25$

Slope and Y-Intercept

Find the slope and y-intercept of each equation.

1) $y = \frac{1}{5}x + 4$

2) $y = 7x + 8$

3) $x - 3y = 9$

4) $y = 5x + 21$

5) $y = 9$

6) $y = -2x + 3$

7) $x = -15$

8) $y = 7x$

9) $y - 3 = 4(x + 1)$

10) $x = -\frac{5}{8}y - \frac{1}{3}$

WWW.MathNotion.com

Slope and One Point

Find a Point-Slope equation for a line containing the given point and having the given slope.

1) $m = -2, (1, -1)$

2) $m = 3, (1, 2)$

3) $m = -2, (-1, -5)$

4) $m = 1, (3, 2)$

5) $m = 5, (2, 4)$

6) $m = \frac{3}{2}, (4, 5)$

7) $m = 0, (-4, -5)$

8) $m = 2, (1, -3)$

9) $m = 1, (0, 3)$

10) $m = \frac{3}{4}, (-2, -5)$

11) $m = -3, (1, -1)$

12) $m = -2, (2, -1)$

13) $m = 5, (1, 0)$

14) $m =$ undefined, $(8, -8)$

15) $m = -\frac{1}{8}, (8, 4)$

16) $m = \frac{1}{4}, (3, 2)$

17) $m = -8, (2, 4)$

18) $m = 6, (-2, -4)$

19) $m = \frac{1}{3}, (3, 1)$

20) $m = \frac{-4}{9}, (0, -3)$

21) $m = \frac{1}{4}, (4, 4)$

22) $m = -5, (0, -1)$

23) $m = 0, (0.9, -3)$

24) $m = -\frac{5}{7}, (7, -1)$

25) $m = 0, (-4, 8)$

26) $m =$ Undefined, $(-10, -2)$

Slope of Two Points

Write the slope-intercept form of the equation of the line through the given points.

1) $(1, 0), (-1, 5)$

2) $(-1, 3), (5, 6)$

3) $(-5, 1), (-1, 5)$

4) $(2, -3), (-9, 8)$

5) $(5, 0), (3, 1)$

6) $(9, -1), (-1, 9)$

7) $(-5, 3), (-6, 1)$

8) $(-7, -2), (1, 0)$

9) $(-5, -5), (3, 3)$

10) $(-1, 9), (-1, -5)$

11) $(-2, 7), (1, 7)$

12) $(1, -5), (4, -4)$

13) $(6, -9), (-3, 0)$

14) $(1, -4), (7, 4)$

15) $(-9, 5), (-3, -1)$

16) $(9, 5), (5, 1)$

17) $(10, -7), (2, -6)$

18) $(-5, -9), (-7, 2)$

19) $(7, 4), (3, 1)$

20) $(-1, -1), (9, 2)$

21) $(-8, 8), (8, 2)$

22) $(9, 2), (5, 11)$

23) $(8, 2), (9, 3)$

24) $(-2, -5), (-5, -8)$

Equation of Parallel and Perpendicular lines

Write the slope-intercept form of the equation of the line described.

1) Through: $(-5, 2)$, parallel to $y = 2x + 5$

2) Through: $(-4, 1)$, parallel to $y = -3x$

3) Through: $(-10, -2)$, perpendecular to $y = \frac{1}{2}x + 8$

4) Through: $(6, -2)$, parallel to $y = -5x + 13$

5) Through: $(-7, 4)$, parallel to $y = \frac{3}{7}x - 6$

6) Through: $(2, 0)$, perpendecular to $y = -\frac{1}{5}x + 8$

7) Through: $(4, -7)$, perpendecular to $y = -6x - 10$

8) Through: $(-5, 1)$, perpendecular to $y = -\frac{1}{8}x + 3$

9) Through: $(-1, -2)$, parallel to $2y + 4x = 9$

10) Through: $(1, 10)$, parallel to $y = \frac{1}{10}x - 5$

11) Through: $(5, -5)$, parallel to $y = 9$

12) Through: $(7, 2)$, perpendecular to $y = \frac{5}{2}x + 3$

13) Through: $(0, -4)$, perpendecular to $3y - x = 11$

14) Through: $(3, 5)$, parallel to $3y + x = 5\frac{3}{4}$

15) Through: $(1, 1)$, perpendecular to $y = 5x + 12$

16) Through: $(-3, -5)$, parallel to $8y - x = 10$

17) Through: $(-2, -2)$, perpendecular to $y = 4x + \frac{1}{7}$

18) Through: $(-8, 0)$, perpendecular to $5y - 4x - 9 = 0$

Quadratic Equations - Quadratic Formula

Solve each equation with the quadratic formula.

1) $3x^2 + 6x - 24 = 0$

2) $4x^2 - 8x = -4$

3) $\frac{1}{4}x^2 = \frac{9}{4}x - 5$

4) $\frac{2}{3}x^2 + \frac{10}{3}x - 4 = 0$

5) $3x^2 = 27x - 60$

6) $3x^2 - 12x - 26 = 10$

7) $7x^2 = -21x + 280$

8) $3x^2 + 15x - 18 = 0$

9) $x^2 + x - 2 = \frac{1}{4}$

10) $\frac{4}{3}x^2 - \frac{2}{3}x = 3$

11) $x^2 = -3x + 40$

12) $9x^2 - 25 = 8x$

13) $\frac{8}{5}x^2 - \frac{8}{5}x = 3$

14) $6x^2 - 3x - 10 = 8$

15) $\frac{1}{3}x^2 = 3x - \frac{100}{15}$

16) $10x^2 + 30x = 400$

17) $x^2 - 2 = \frac{1}{8}x$

18) $24x^2 + 18x + 15 = 0$

19) $x^2 - \frac{1}{2}x - \frac{13}{2} = 1$

20) $11x^2 + 1 = 5x^2 + 7x$

21) $17x^2 + 14 = x$

22) $10x^2 - 5x - 20 = 10$

Answer key Chapter 7

Relation and Functions

1) No, $D_f = \{4, -7, 1, -5, 15\}$, $R_f = \{-6, 2, 4, -9, 10\}$
2) Yes, $D_f = \{4, 2, -6, 6, -11\}$, $R_f = \{5, 3, -8, 2\}$
3) Yes, $D_f = (-\infty, \infty)$, $R_f = \{-2, \infty)$
4) No, $D_f = \{2, 7, 9, 8, 7\}$, $R_f = \{-2, -6, 9, 1, 4\}$
5) No, $D_f = [-3, 2]$, $R_f = [-2, 3]$
6) Yes, $D_f = \{9, 11, 5, -3\}$, $R_f = \{2, 8, -3, 17\}$

Slope form

1) $y = -\frac{4}{5}x + 3$
2) $y = -\frac{1}{3}x + \frac{1}{4}$
3) $y = -7x - 9$
4) $y = \frac{7}{11}x + \frac{5}{11}$
5) $y = \frac{5}{4}x - \frac{7}{4}$
6) $y = 7x + 2$
7) $y = -2x$
8) $y = \frac{5}{7}x + \frac{9}{7}$
9) $y = 0.5x + 2$
10) $y = 4.5x + 27$
11) $y = -10x - 120$
12) $y = -\frac{1}{12}x - \frac{3}{4}$
13) $y = 0.5x - 1$
14) $y = -4x + 5$

Slope and Y-Intercept

1) $m = \frac{1}{5}, b = 4$
2) $m = 7, b = 8$
3) $m = \frac{1}{3}, b = -3$
4) $m = 5, b = 21$
5) $m = 0, b = 9$
6) $m = -2, b = 3$
7) $m = undefind$, $b:$ no intercept
8) $m = 7, b = 0$
9) $m = 4, b = 7$
10) $m = -\frac{8}{5}, b = -\frac{1}{3}$

Slope and One Point

1) $y = -2x + 1$
2) $y = 3x - 1$
3) $y = -2x - 7$
4) $y = x - 1$
5) $y = 5x - 6$
6) $y = \frac{3}{2}x - 1$
7) $y = -5$
8) $y = 2x - 5$
9) $y = x + 3$
10) $y = \frac{3}{4}x - \frac{7}{2}$
11) $y = -3x + 2$
12) $y = -2x + 3$
13) $y = 5x$
14) $x = 8$
15) $y = -\frac{1}{8}x + 5$
16) $y = \frac{1}{4}x + \frac{5}{4}$
17) $y = -8x + 20$
18) $y = 6x + 8$
19) $y = \frac{1}{3}x$
20) $y = -\frac{4}{9}x - 3$
21) $y = \frac{1}{4}x + 3$
22) $y = -5x - 1$

Common Core Math Practice Grade 7

23) $y = -3$

24) $y = -\frac{5}{7}x + 4$

25) $y = 8$

26) $x = -10$

Slope of Two Points

1) $y = -\frac{5}{2}x + \frac{5}{2}$

2) $y = \frac{1}{2}x + \frac{7}{2}$

3) $y = x + 6$

4) $y = -x - 1$

5) $y = -\frac{1}{2}x + \frac{5}{2}$

6) $y = -x + 8$

7) $y = 2x + 13$

8) $y = \frac{1}{4}x - \frac{1}{4}$

9) $y = x$

10) $x = -1$

11) $y = 7$

12) $y = \frac{1}{3}x - 5\frac{1}{3}$

13) $y = -x - 3$

14) $y = \frac{4}{3}x - 5\frac{1}{3}$

15) $y = -x - 4$

16) $y = x - 4$

17) $y = -\frac{1}{8}x - 5\frac{3}{4}$

18) $y = -5\frac{1}{2}x - 36\frac{1}{2}$

19) $y = \frac{3}{4}x - 1\frac{1}{4}$

20) $y = \frac{3}{10}x - \frac{7}{10}$

21) $y = -\frac{3}{8}x + 5$

22) $y = -\frac{9}{4}x + 22\frac{1}{4}$

23) $y = x - 6$

24) $y = x - 3$

Equation of Parallel and Perpendicular lines

1) $y = 2x + 12$

2) $y = -3x - 11$

3) $y = -2x - 22$

4) $y = -5x + 28$

5) $y = \frac{3}{7}x + 7$

6) $y = 5x - 10$

7) $y = \frac{1}{6}x - 7\frac{2}{3}$

8) $y = 8x + 41$

9) $y = -2x - 4$

10) $y = \frac{1}{10}x + 9\frac{9}{10}$

11) $y = -5$

12) $y = -\frac{2}{5}x + 4\frac{4}{5}$

13) $y = -3x - 4$

14) $y = -\frac{1}{3}x + 6$

15) $y = -\frac{1}{5}x + 1\frac{1}{5}$

16) $y = \frac{1}{8}x - 4\frac{5}{8}$

17) $y = -\frac{1}{4}x - 2\frac{1}{2}$

18) $y = -\frac{5}{4}x - 10$

Quadratic Equations - Quadratic Formula

1) $\{2, -4\}$

2) $\{1\}$

3) $\{5, 4\}$

4) $\{1, -6\}$

5) $\{5, 4\}$

6) $\{6, -2\}$

7) $\{5, -8\}$

8) $\{1, -6\}$

9) $\{\frac{-1+\sqrt{10}}{2}, \frac{-1-\sqrt{10}}{2}\}$

10) $\{\frac{1+\sqrt{37}}{4}, \frac{1-\sqrt{37}}{4}\}$

11) $\{5, -8\}$

12) $\{\frac{4+\sqrt{241}}{9}, \frac{4-\sqrt{241}}{9}\}$

13) $\{\frac{2+\sqrt{34}}{4}, \frac{2-\sqrt{34}}{4}\}$

14) $\{2, -\frac{3}{2}\}$

15) $\{5, 4\}$

16) $\{5, -8\}$

17) $\{\frac{1+3\sqrt{57}}{16}, \frac{1-3\sqrt{57}}{16}\}$

18) $\{\frac{-3+i\sqrt{31}}{8}, \frac{-3-i\sqrt{31}}{8}\}$

19) $\{3, -\frac{5}{2}\}$

20) $\{1, \frac{1}{6}\}$

21) $\{\frac{1+i\sqrt{951}}{34}, \frac{1-i\sqrt{951}}{34}\}$

22) $\{2, -\frac{3}{2}\}$

Chapter 8:
Equations and Inequality

Distributive and Simplifying Expressions

Simplify each expression.

1) $6x + 2 - 8 =$

2) $-(-4 - 5x) =$

3) $(-3x + 4)(-2) =$

4) $(-2x)(x + 3) =$

5) $-2x + x^2 + 4x^2 =$

6) $7y + 7x + 8y - 5x =$

7) $-3x + 3y + 14x - 9y =$

8) $-2x - 5 + 8x + \frac{16}{4} =$

9) $5 - 8(x - 2) =$

10) $-5 - 5x + 3x =$

11) $(x - 3y)2 + 4y =$

12) $2.5x^2 \times (-5x) =$

13) $-4 - 2x^2 + 6x^2 =$

14) $8 + 14x^2 + 4 =$

15) $4(-2x - 7) + 10 =$

16) $(-x)(-2 + 3x) - x(7 + x) =$

17) $-3(6 + 12) - 3x + 5x =$

18) $-4(5 - 12x - 3x) =$

19) $3(-2x - 6) =$

20) $9 + 7x - 9 =$

21) $x(-2x + 8) =$

22) $5xy + 4x - 3y + x + 2y =$

23) $3(-x - 7) + 9 =$

24) $(-3x - 4) + 7 =$

25) $3x + 4y - 5 + 1 =$

26) $(-2 + 3x) - 3x(1 + 2x) =$

27) $(-3)(-3x - 3y) =$

28) $4(-x - 2) + 5 =$

Factoring Expressions

Factor the common factor out of each expression.

1) $12x - 6 =$

2) $5x - 15 =$

3) $\frac{45}{15}x - 15 =$

4) $7b - 28 =$

5) $4a^2 - 24a =$

6) $2xy - 10y =$

7) $5x^2y + 15x =$

8) $a^2 - 8a + 7ab =$

9) $2a^2 + 2ab =$

10) $4x + 20 =$

11) $24x - 36xy =$

12) $8x - 6 =$

13) $\frac{1}{4}x - \frac{3}{4}y =$

14) $7xy - \frac{14}{3}x =$

15) $3ab + 9c =$

16) $\frac{1}{3}x - \frac{4}{3} =$

17) $10x - 15xy =$

18) $x^2 + 8x =$

19) $4x^2 - 12y =$

20) $4x^3 + 3xy + x^2 =$

21) $21x - 14 =$

22) $20b - 60c + 20d =$

23) $24ab - 8ac =$

24) $ax - ay - 3x + 3y =$

25) $3ax + 4a + 9x + 12 =$

26) $x^2 - 10x =$

27) $9x^3 - 18x^2 =$

28) $5x^2 - 70xy =$

Evaluate One Variable Expressions

Evaluate each using the values given.

1) $x + 4x, x = 3$

2) $5(-6 + 3x), x = 1$

3) $4x + 7x, x = -3$

4) $5(2 - x) + 5, x = 3$

5) $6x + 4x - 10, x = 2$

6) $5x + 11x + 12, x = -1$

7) $5x - 2x - 4, x = 5$

8) $\frac{3(5x+8)}{9}, x = 2$

9) $2x - 85, x = 32$

10) $\frac{x}{18}, x = 108$

11) $7(3 + 2x) - 33, x = 5$

12) $7(x + 3) - 23, x = 4$

13) $\frac{x+(-6)}{-3}, x = -6$

14) $8(6 - 3x) + 5, x = 2$

15) $-11 - \frac{x}{5} + 3x, x = 10$

16) $5x + 11x, x = 1$

17) $-12x + 3(5 + 3x), x = -7$

18) $x + 11x, x = 0.5$

19) $\frac{(2x-2)}{6}, x = 13$

20) $3(-1 - 2x), x = 5$

21) $5x - (5 - x), x = 3$

22) $\left(-\frac{15}{x}\right) + 2 + x, x = 5$

23) $-\frac{x \times 5}{x}, x = 5$

24) $2(-1 - 3x), x = 2$

25) $2x^2 + 7x, x = 1$

26) $2(3x + 1) - 4(x - 5), x = 3$

27) $-6x - 4, x = -5$

28) $7x + 2x, x = 3$

Common Core Math Practice Grade 7

Evaluate Two Variable Expressions

Evaluate the expressions.

1) $x + 4y$, $x = 5, y = 2$

2) $(-2)(-3x - 2y)$, $x = 1, y = 2$

3) $4x + 2y$, $x = 10, y = 5$

4) $\frac{x-4}{y+1}$, $x = 8, y = 3$

5) $\frac{a}{4} - 6b$, $a = 32, b = 4$

6) $3x - 4(y - 8)$, $x = 5, y = 3$

7) $3x + 2y - 10$, $x = 2, y = 10$

8) $-3x + 10 + 8y - 5$, $x = 2, y = 1$

9) $yx \div 3$, $x = 9, y = 9$

10) $a - b \div 3$, $a = 3, b = 12$

11) $6(x - y)$, $x = 7, y = 4$

12) $5x - 4y$, $x = 5, y = 8$

13) $\frac{10}{a} + 3b$, $a = 5, b = 4$

14) $2x^2 + 4xy$, $x = 3, y = 5$

15) $8 - \frac{xy}{10} + y$, $x = 6, y = 5$

16) $7(3x - y)$, $x = 7, y = -9$

17) $5x^2 - 3y^2$, $x = -1, y = 2$

18) $3x + \frac{y}{4}$, $x = 6, y = 16$

19) $4(4x - 2y)$, $x = 3, y = 5$

20) $4x(y - \frac{1}{2})$, $x = 5, y = 4$

21) $5(x^2 - 2y)$, $x = 3, y = 2$

22) $5xy$, $x = 2, y = 8$

23) $\frac{1}{3}y^3\left(y - \frac{1}{4}x\right)$, $x = -4, y = 3$

24) $-3(x - 5y) - 2x$, $x = 4, y = 2$

25) $-2x + \frac{1}{6}xy$, $x = 3, y = 6$

26) $x^2 + xy^2$, $x = 5, y = 7$

27) $x - 2y + 8$, $x = 9, y = 6$

28) $\frac{xy}{2x+y}$, $x = 5, y = 4$

www.MathNotion.com

Graphing Linear Equation

Sketch the graph of each line.

1) $y = 2x - 5$
2) $y = -2x + 3$
3) $x - y = 0$

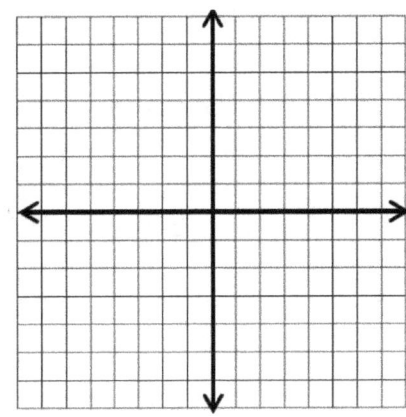

4) $x + y = 3$
5) $5x + 3y = -2$
6) $y - 3x + 2 = 0$

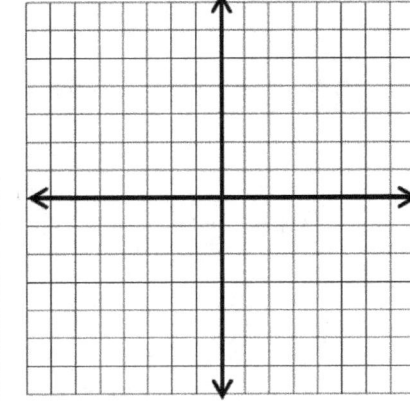

One Step Equations

Solve each equation.

1) $44 = (-12) + x$

2) $8x = (-64)$

3) $(-72) = (-8x)$

4) $(-5) = 3 + x$

5) $4 + \frac{x}{2} = (-3)$

6) $8x = (-104)$

7) $62 = x - 13$

8) $\frac{x}{3} = (-15)$

9) $x + 112 = 154$

10) $x - \frac{1}{3} = \frac{2}{3}$

11) $(-24) = x - 32$

12) $(-3x) = 39$

13) $(-169) = (13x)$

14) $-4x + 42 = 50$

15) $5x + 3 = 38$

16) $80 = (-8x)$

17) $3x + 7 = 19$

18) $24x = 144$

19) $x - 18 = 15$

20) $0.9x = 4.5$

21) $4x = 84$

22) $2x + 2.98 = 66.98$

23) $x + 9 = 6$

24) $x + 14 = 6$

25) $9x + 41 = 5$

26) $\frac{1}{4}x + 30 = 12$

Two Steps Equations

Solve each equation.

1) $6(3 + x) = 42$

2) $(-7)(x - 2) = 56$

3) $(-8)(3x - 4) = (-16)$

4) $5(2 + x) = -15$

5) $19(3x + 11) = 38$

6) $4(2x + 2) = 24$

7) $5(8 + 3x) = (-20)$

8) $(-5)(5x - 3) = 40$

9) $2x + 12 = 16$

10) $\frac{4x - 5}{5} = 3$

11) $(-3) = \frac{x + 4}{7}$

12) $80 = (-8)(x - 3)$

13) $\frac{x}{3} + 7 = 19$

14) $\frac{1}{4} = \frac{1}{2} + \frac{x}{4}$

15) $\frac{11 + x}{5} = (-6)$

16) $(-3)(10 + 5x) = (-15)$

17) $(-3x) + 12 = 24$

18) $\frac{x + 5}{5} = -5$

19) $\frac{x + 23}{8} = 3$

20) $(-4) + \frac{x}{2} = (-14)$

21) $-5 = \frac{x + 7}{8}$

22) $\frac{9x - 3}{6} = 4$

23) $\frac{2x - 12}{8} = 6$

24) $40 = (-5)(x - 8)$

Multi Steps Equations

Solve each equation.

1) $2 - (4 - 5x) = 3$

2) $-15 = -(4x + 7)$

3) $6x - 18 = (-2x) + 6$

4) $-32 = (-5x) - 11x$

5) $3(2 + 3x) + 3x = -30$

6) $5x - 18 = 2 + 2x - 7 + 2x$

7) $12 - 6x = (-36) - 3x + 3x$

8) $16 - 4x - 4x = 8 - 4x$

9) $8 + 7x + x = (-12) + 3x$

10) $(-3x) - 3(-2 + 4x) = 366$

11) $20 = (-200x) - 5 + 5$

12) $61 = 5x - 23 + 7x$

13) $7(4 + 2x) = 140$

14) $-60 = (-7x) - 13x$

15) $2(4x + 5) = -2(x + 4) - 22$

16) $11x - 17 = 6x + 8$

17) $9 = -3(x - 8)$

18) $(-6) - 8x = 6(1 + 2x)$

19) $x + 3 = -2(9 + 3x)$

20) $10 = 4 - 5x - 9$

21) $-15 - 9x - 3x = 12 - 3x$

22) $-23 - 3x + 5x = 27 - 23x$

23) $19 - 6x - 9x = -5 - 9x$

24) $15x - 18 = 6x + 9$

Graphing Linear Inequalities

Sketch the graph of each linear inequality.

1) $y > 2x - 3$ 2) $y < x + 3$ 3) $y \leq -3x - 8$

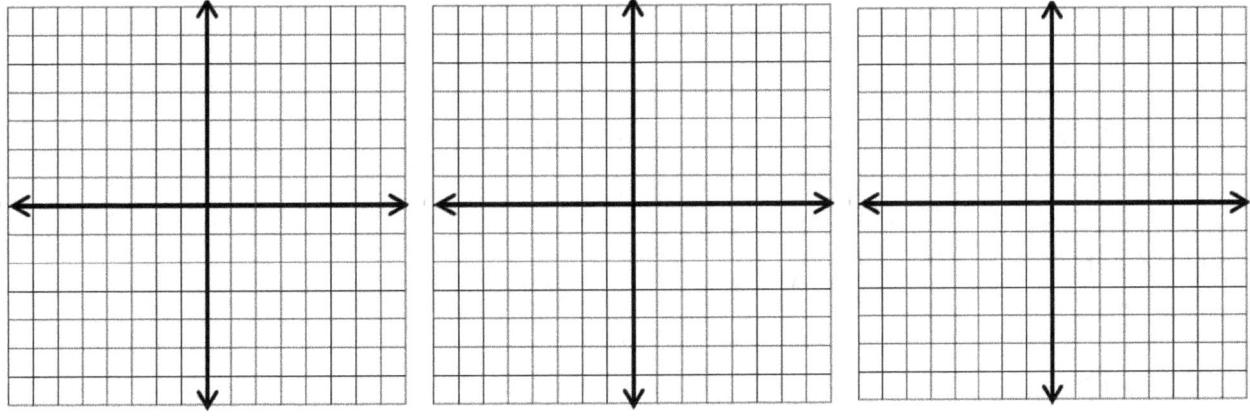

4) $3y \geq 6 + 3x$ 5) $-3y < x - 12$ 6) $2y \geq -8x + 4$

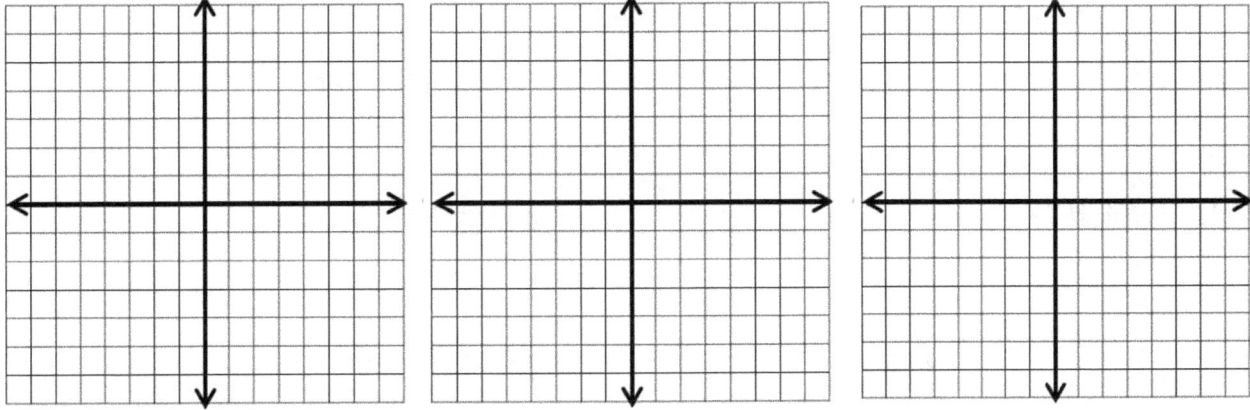

One Step Inequality

Solve each inequality.

1) $7x < 14$

2) $x + 7 \geq -8$

3) $x - 1 \leq 9$

4) $-2x + 4 > -10$

5) $x + 18 \geq -6$

6) $x + 9 \geq 5$

7) $x - \frac{1}{3} \leq 5$

8) $-7x < 42$

9) $-x + 8 > -3$

10) $\frac{x}{3} + 3 > -9$

11) $-x + 8 > -4$

12) $x - 14 \leq 18$

13) $-x - 5 \leq -7$

14) $x + 26 \geq -13$

15) $x + \frac{1}{3} \geq -\frac{2}{3}$

16) $x + 6 \geq -14$

17) $x - 42 \leq -48$

18) $x - 5 \leq 4$

19) $-x + 5 > -6$

20) $x + 6 \geq -12$

21) $8x + 6 \leq 22$

22) $4x - 3 \geq 9$

23) $3x - 5 < 22$

24) $6x - 8 \leq 40$

Two Steps Inequality

Solve each inequality

1) $2x - 3 \leq 7$

2) $3x - 4 \leq 8$

3) $\frac{-1}{4}x + \frac{x}{2} \leq \frac{1}{8}$

4) $5x + 10 \geq 30$

5) $4x - 7 \geq 9$

6) $3x - 5 \leq 16$

7) $8x - 2 \leq 14$

8) $9x + 5 \leq 23$

9) $2x + 10 > 32$

10) $\frac{x}{8} + 2 \leq 4$

11) $3x + 4 \geq 37$

12) $3x - 8 < 10$

13) $6 \geq \frac{x+7}{2}$

14) $3x + 9 < 48$

15) $\frac{4+x}{5} \geq 3$

16) $16 + 4x < 36$

17) $16 > 6x - 8$

18) $5 + \frac{x}{3} < 6$

19) $-4 + 4x > 24$

20) $5 + \frac{x}{7} < 3$

Multi Steps Inequality

Solve the inequalities.

1) $4x - 6 < 5x - 9$

2) $\frac{4x + 5}{3} \leq x$

3) $7x - 5 > 3x + 15$

4) $-3x > -6x + 4$

5) $3 + \frac{x}{2} < \frac{x}{4}$

6) $\frac{4x - 6}{8} > x$

7) $4x - 20 + 4 > 6x - 8$

8) $x - 8 > 11 + 3(x + 5)$

9) $\frac{x}{3} + 2 > x$

10) $-7x + 8 \geq -6(4x - 8) - 8x$

11) $7x - 4 \leq 8x + 9$

12) $\frac{2x - 7}{5} > 2$

13) $8(x + 2) < 6x + 10$

14) $-8x + 12 \leq 4(x - 9)$

15) $\frac{5x - 6}{3} > 3x + 2$

16) $2(x - 8) + 10 \geq 4x - 2$

17) $\frac{-5x + 7}{6} > 5x$

18) $-3x - 4 > -7x$

19) $\frac{1}{4}x - 12 > \frac{1}{8}x - 19$

20) $-4(x - 9) \leq 5x$

Common Core Math Practice Grade 7

Finding Distance of Two Points

Find the distance between each pair of points.

1) $(2, 1), (-1, -3)$
2) $(-4, -2), (4, 4)$
3) $(-3, 0), (15, 24)$
4) $(-4, -1), (1, 11)$
5) $(3, -2), (-6, -14)$
6) $(-6, 0), (-2, 3)$
7) $(3, 2), (11, 17)$
8) $(-6, -10), (6, -1)$
9) $(5, 9), (-11, -3)$
10) $(6, -2), (2, -6)$
11) $(3, 0), (18, 36)$
12) $(8, 4), (3, -8)$
13) $(4, 2), (-5, -10)$
14) $(-8, 10), (4, 40)$
15) $(8, 4), (-10, -20)$
16) $(-8, -2), (16, 8)$
17) $(3, 5), (-5, -10)$
18) $(-10, 20), (35, 45)$

Find the midpoint of the line segment with the given endpoints.

1) $(-2, -2), (4, 2)$
2) $(10, 4), (-2, 4)$
3) $(12, -2), (4, 10)$
4) $(-6, -5), (2, 1)$
5) $(3, -2), (5, -2)$
6) $(-10, -4), (6, -2)$
7) $(4, 1), (-4, 9)$
8) $(-5, 6), (-5, 2)$
9) $(-8, 8), (4, -2)$
10) $(1, 7), (5, -1)$
11) $(-9, 5), (5, 3)$
12) $(7, 10), (-3, -6)$
13) $(-8, 14), (-8, 2)$
14) $(16, 7), (6, -3)$
15) $(5, 6), (-3, 4)$
16) $(-9, -1), (-5, 7)$
17) $(17, 9), (5, 11)$
18) $(-8, -11), (18, -1)$

www.MathNotion.com

Answer key Chapter 8

Distributive and Simplifying Expressions

1) $6x - 6$
2) $4 + 5x$
3) $6x - 8$
4) $-2x^2 - 6x$
5) $5x^2 - 2x$
6) $2x + 15y$
7) $11x - 6y$
8) $6x - 1$
9) $-8x + 21$
10) $-2x - 5$
11) $2x - 2y$
12) $-12.5x^3$
13) $4x^2 - 4$
14) $14x^2 + 12$
15) $-8x - 18$
16) $-4x^2 - 5x$
17) $2x - 54$
18) $60x - 20$
19) $-6x - 18$
20) $7x$
21) $-2x^2 + 8x$
22) $5x + y + 5xy$
23) $-3x - 12$
24) $-3x + 3$
25) $3x + 4y - 4$
26) $-6x^2 - 2$
27) $9x + 9y$
28) $-4x - 3$

Factoring Expressions

1) $3(4x - 2)$
2) $5(x - 3)$
3) $3(x - 5)$
4) $7(b - 4)$
5) $4a(a - 6)$
6) $2y(x - 5)$
7) $5x(xy + 3)$
8) $a(a - 8 + 7b)$
9) $2a(a + b)$
10) $4(x + 5)$
11) $12x(2 - 3y)$
12) $2(4x - 3)$
13) $\frac{1}{4}(x - 3y)$
14) $7x(y - \frac{2}{3})$
15) $3(ab + 3c)$
16) $\frac{1}{3}(x - 4)$
17) $5x(2 - 3y)$
18) $x(x + 8)$
19) $4(x^2 - 3y)$
20) $x(4x^2 + 3y + x)$
21) $7(3x - 2)$
22) $20(b - 3c + d)$
23) $8a(3b - c)$
24) $(x - y)(a - 3)$
25) $(3x + 4)(a + 3)$
26) $x(x - 10)$
27) $9x^2(x - 2)$
28) $5x(x - 14y)$

Evaluate One Variable Expressions

1) 15
2) −15
3) −33
4) 0
5) 10
6) −4
7) 11
8) 6
9) −21
10) 6
11) 58
12) 26
13) 4
14) 5
15) 17
16) 16
17) 36
18) 6
19) 4
20) −33
21) 13
22) 4
23) −5
24) −14

25) 9   26) 28   27) 26   28) 27

Evaluate Two Variable Expressions

1) 13   9) 27   17) −7   24) 10
2) 14   10) 3   18) 22   25) −3
3) 50   11) 18   19) 8   26) 270
4) 1   12) 20   20) 70   27) 5
5) −16   13) 17   21) 25   28) $\frac{10}{7}$
6) 35   14) 78   22) 80
7) 16   15) 10   23) 36
8) 7   16) 210

Graphing Lines Using Line Equation

1) $y = 2x - 5$ 2) $y = -2x + 3$ 3) $x - y = 0$

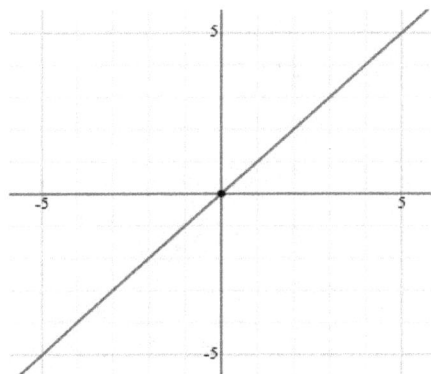

4) $x + y = 3$ 5) $5x + 3y = -2$ 6) $y - 3x + 2 = 0$

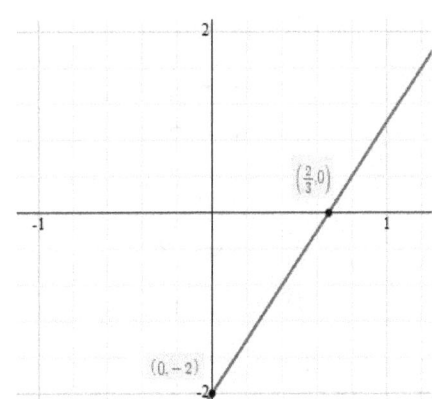

Common Core Math Practice Grade 7

One Step Equations

1) $x = 56$
2) $x = -8$
3) $x = 9$
4) $x = -8$
5) $x = -14$
6) $x = -13$
7) $x = 75$
8) $x = -45$
9) $x = 42$
10) $x = 1$
11) $x = 8$
12) $x = -13$
13) $x = -13$
14) $x = -2$
15) $x = 7$
16) $x = -10$
17) $x = 4$
18) $x = 6$
19) $x = 33$
20) $x = 5$
21) $x = 21$
22) $x = 32$
23) $x = -3$
24)
25)
26)

Two Steps Equations

1) $x = 4$
2) $x = -6$
3) $x = 2$
4) $x = -5$
5) $x = -3$
6) $x = 2$
7) $x = -4$
8) $x = -1$
9) $x = 2$
10) $x = 5$
11) $x = -25$
12) $x = -7$
13) $x = 36$
14) $x = -1$
15) $x = -41$
16) $x = -1$
17) $x = -4$
18) $x = -30$
19) $x = 1$
20) $x = -20$
21) $x = -47$
22) $x = 3$
23) $x = 30$
24) $x = 0$

Multi Steps Equations

1) $x = 1$
2) $x = 2$
3) $x = 3$
4) $x = 2$
5) $x = -3$
6) $x = 13$
7) $x = 8$
8) $x = 2$
9) $x = -4$
10) $x = -24$
11) $x = -0.1$
12) $x = 7$
13) $x = 8$
14) $x = 3$
15) $x = -4$
16) $x = 5$
17) $x = 5$
18) $x = -3/5$
19) $x = -3$
20) $x = -3$
21) $x = -3$
22) $x = 2$
23) $x = 4$
24) $x = 3$

WWW.MathNotion.com

Common Core Math Practice Grade 7

Graphing Linear Inequalities

1) $y > 2x - 3$
2) $y < x + 3$
3) $y \leq -3x - 8$

4) $3y \geq 6 + 3x$
5) $-3y < x - 12$
6) $2y \geq -8x + 4$

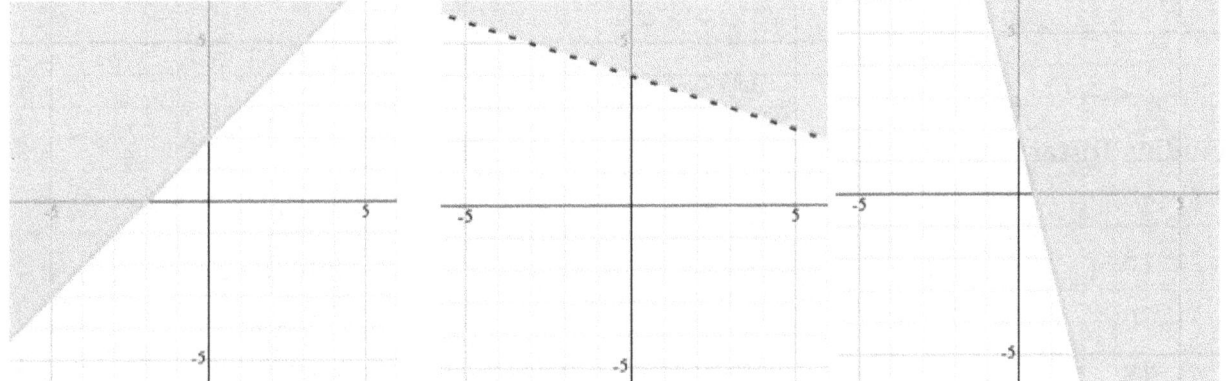

One Step Inequality

1) $x < 2$
2) $x \geq -15$
3) $x \leq 10$
4) $x < 7$
5) $x \geq -24$
6) $x \geq -4$
7) $x \leq \frac{16}{3}$
8) $x > -6$
9) $x < 11$
10) $x > -36$
11) $x < 12$
12) $x \leq 32$
13) $x \geq 2$
14) $x \geq -39$
15) $x \geq -1$
16) $x \geq -20$
17) $x \leq -6$
18) $x \leq 9$
19) $x < 11$
20) $x \geq -18$
21) $x \leq 2$
22) $x \geq 3$
23) $x < 9$
24) $x \leq 8$

Two Steps Inequality

1) $x \leq 5$
2) $x \leq 4$
3) $x \leq 0.5$
4) $x \geq 4$
5) $x \geq 4$
6) $x \leq 7$

WWW.MathNotion.com

7) $x \leq 2$
8) $x \leq 2$
9) $x > 11$
10) $x \leq 16$
11) $x \geq 11$

12) $x < 6$
13) $x \leq 5$
14) $x < 13$
15) $x \geq 11$
16) $x < 5$

17) $x < 4$
18) $x < 3$
19) $x > 7$
20) $x < -14$

Multi Steps Inequality

1) $x > 3$
2) $x \leq -5$
3) $x > 5$
4) $x > \frac{4}{3}$
5) $x < -12$
6) $x < -1.5$
7) $x < -4$

8) $x < -17$
9) $x < 3$
10) $x \geq 1.6$
11) $x \geq -13$
12) $x > 8.5$
13) $x < -3$
14) $x \geq 4$

15) $x < -3$
16) $x \leq -2$
17) $x < \frac{1}{5}$
18) $x > 1$
19) $x > -56$
20) $x \geq 4$

Finding Distance of Two Points

1) 5
2) 10
3) 30
4) 13
5) 15
6) 5

7) 17
8) 15
9) 20
10) $4\sqrt{2}$
11) 39
12) 13

13) 15
14) $6\sqrt{29}$
15) 30
16) 26
17) 17
18) $5\sqrt{106}$

Finding Midpoint

1) $(1, 0)$
2) $(4, 4)$
3) $(8, 4)$
4) $(-2, -2)$
5) $(4, -2)$
6) $(-2, -3)$

7) $(0, 5)$
8) $(-5, 4)$
9) $(-2, 3)$
10) $(3, 3)$
11) $(-2, 4)$
12) $(2, 2)$

13) $(-8, 8)$
14) $(11, 2)$
15) $(1, 5)$
16) $(-7, 3)$
17) $(11, 10)$
18) $(5, -6)$

Chapter 9:
Transformations

Translations

Graph the image of the figure using the transformation given.

1) translation: 4 units right and 3 units down

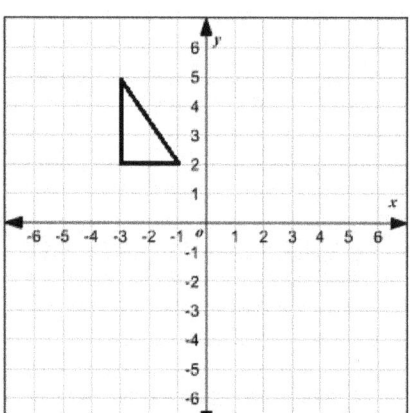

2) translation: 2 units left and 1 units down

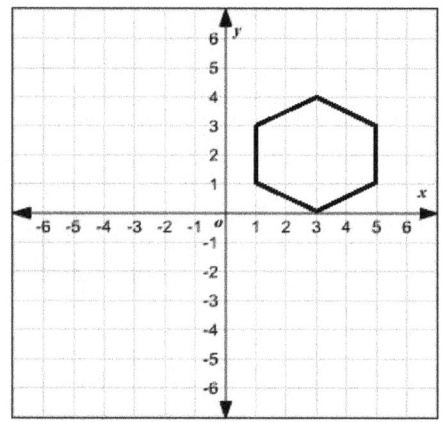

Write a rule to describe each transformation.

3)

4)

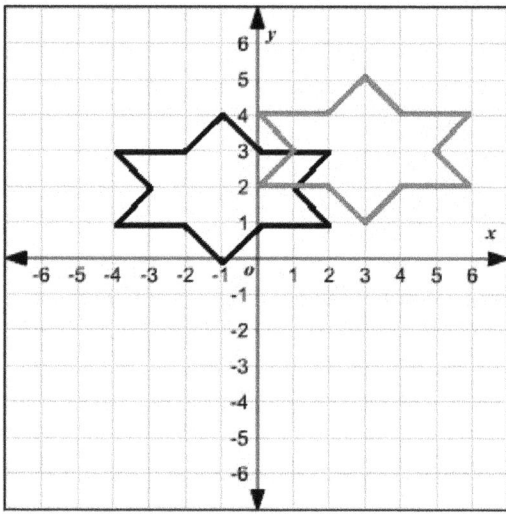

Reflections

Graph the image of the figure using the transformation given.

1) Reflection across $x = -2$

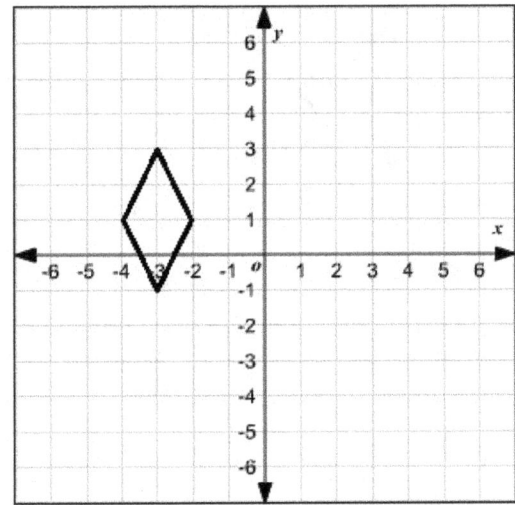

2) Reflection across $y = -x$

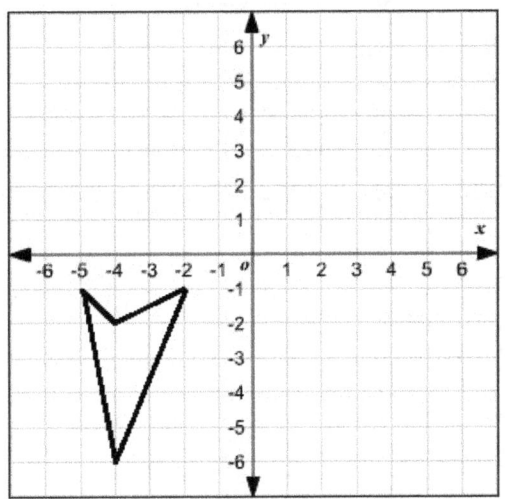

3) Reflection across $y = -1$

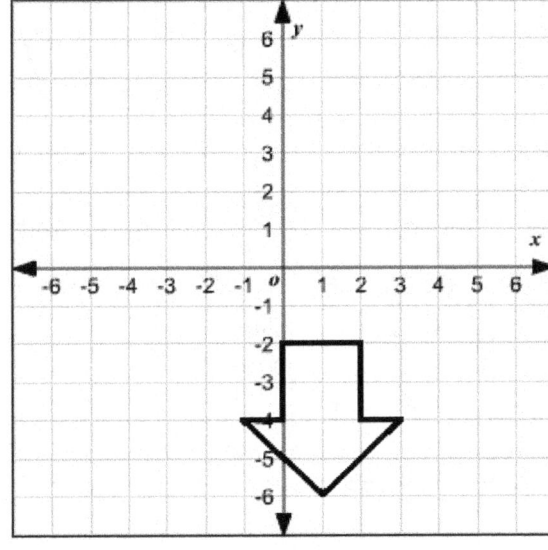

4) Reflection across x axis

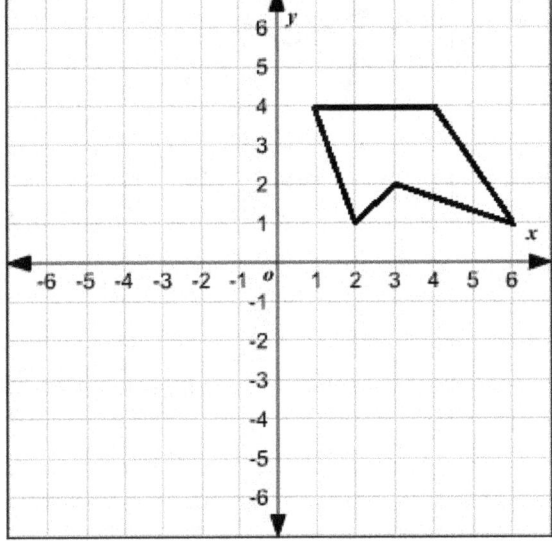

Common Core Math Practice Grade 7

Write a rule to describe each transformation.

5)

6)

7)

8)
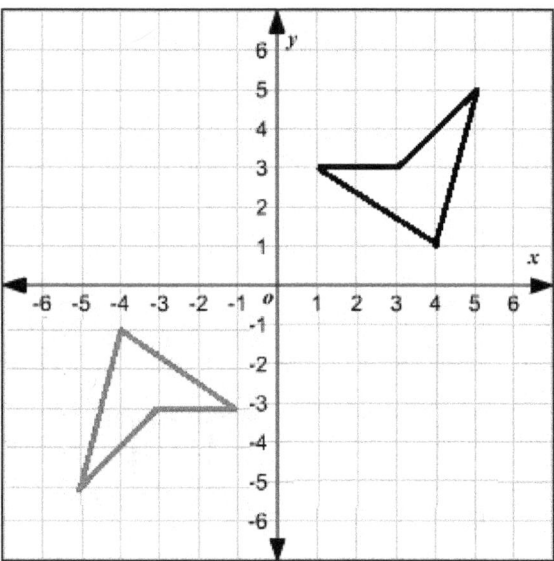

Rotations

Graph the image of the figure using the transformation given.

1) rotation 270° clockwise about the origin

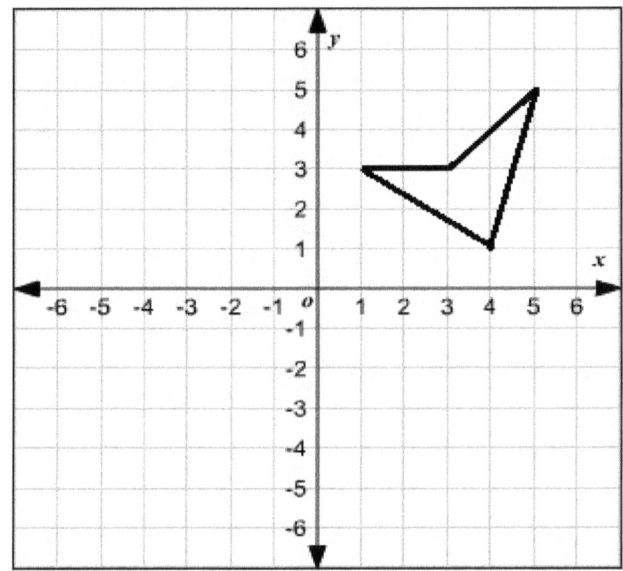

2) rotation 180° clockwise about the origin

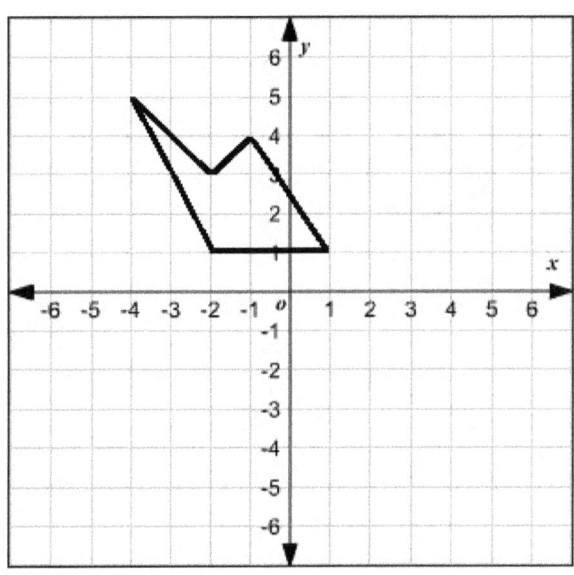

3) rotation 90° clockwise about the origin

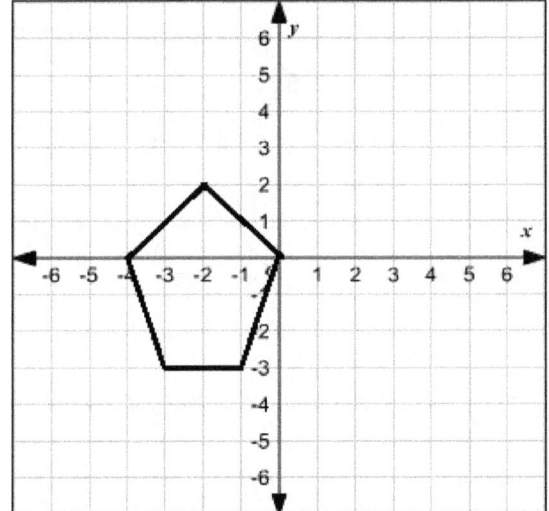

4) rotation 90° counterclockwise about the origin

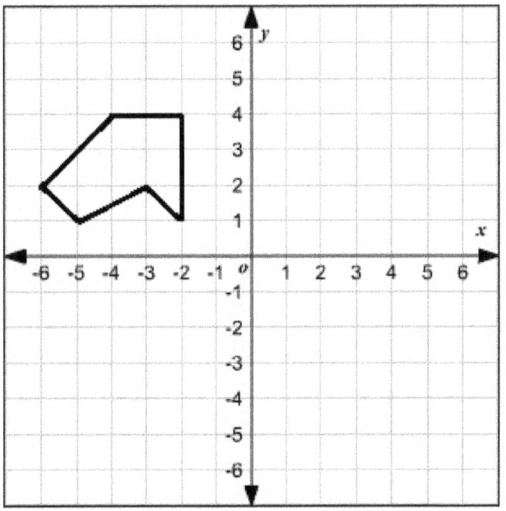

Common Core Math Practice Grade 7

Write a rule to describe each transformation.

5)

6)

7)

8)

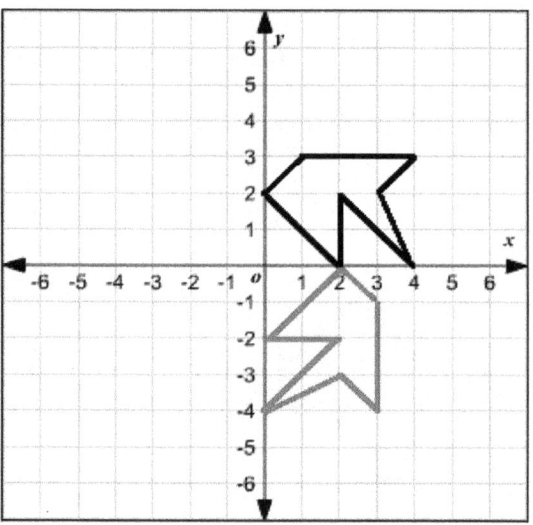

WWW.MathNotion.com

Dilations

Draw a dilation of the figure using the given scale factor.

1) $k = \frac{1}{3}$

2) $k = 2$

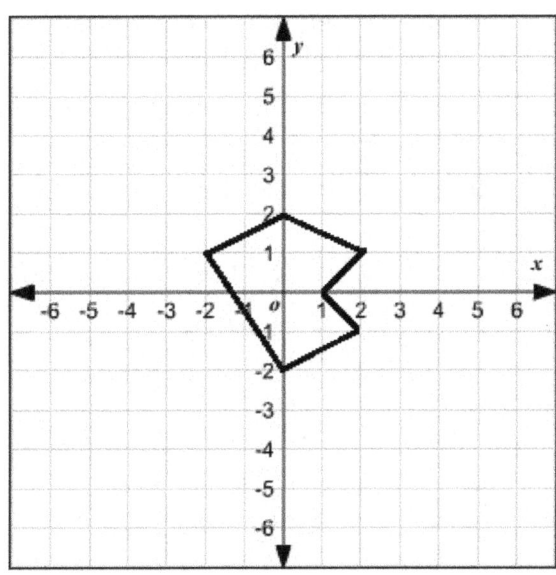

Determine whether the dilation from figure M to figure N is a reduction or an enlargement. Then find the scale factor and the missing length.

3)

4)

Coordinates of Vertices

Calculate the new coordinates after the given transformations.

1) Translate: 1-unit right and 3 units down.

   A(−3, 1), B(−1, −2), C(3, 0)

2) Rotation: 180° clockwise about the origin.

   D(−2, 2), E(−6, 10), F(−8, 4), G(−4, 12)

3) Rotation: 90° counterclockwise about the origin.

   P(1, 2), Q(4, 1), R(5, 5), S(0, −2)

4) Rotation: 270° counterclockwise about the origin.

   J(−3, 4), K(−6, 2), L(1, −3)

5) Reflection: over the x axis.

   C(−1, −4), D(−5, −2), W(2, −6), Y(8, 3)

6) Reflection: across the line $y = x$.

   A(4, −1), B(6, −3), C(5, −5), D(3, −4)

7) Reflection: across the line $y = -3$.

   K(−2, −1), L(−6, 0), M(−3, −2), N(−8, 1)

8) Dilate: Reduction by scale factor $\frac{1}{4}$.

   A(8, 2), B(−10, −4), C(−12, 8)

9) Dilate: Enlargement by scale factor 3.

   F(−2, 0), G(−1, 2), H(2, 2)

Answers of Worksheets – Chapter 9

Translations

1)

2)
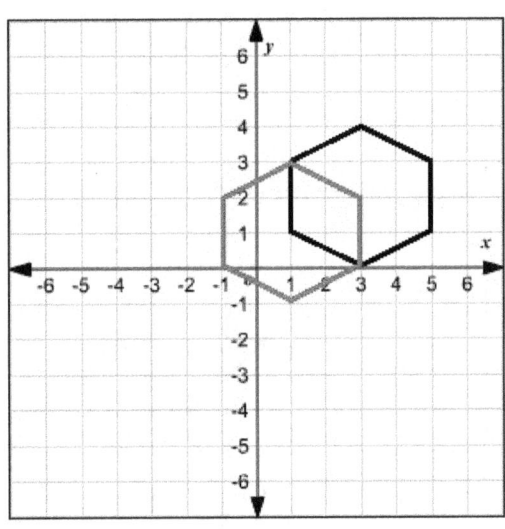

3) translation: : 3 units left and 3 units down

4) translation: 4 units right and 1 unit up

Reflections

1)

2)
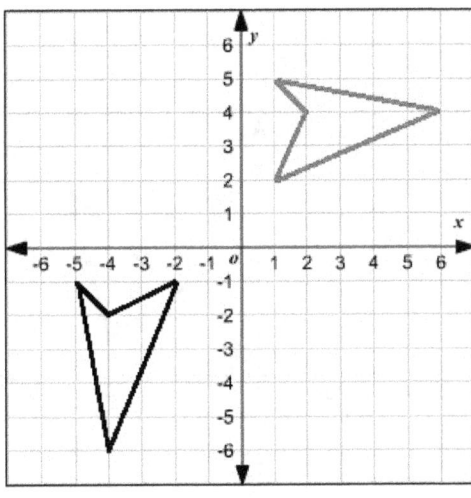

Common Core Math Practice Grade 7

3)

4)

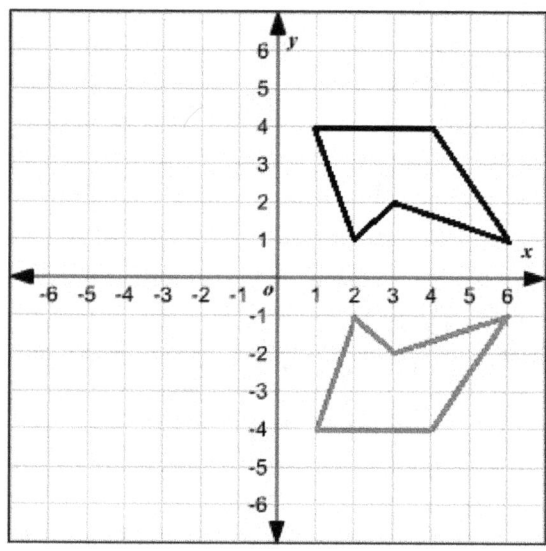

5) reflection across the y = 0 (x axis)

6) reflection across the y = x

7) reflection against the origin

8) reflection against the origin

Rotations

1)

2)

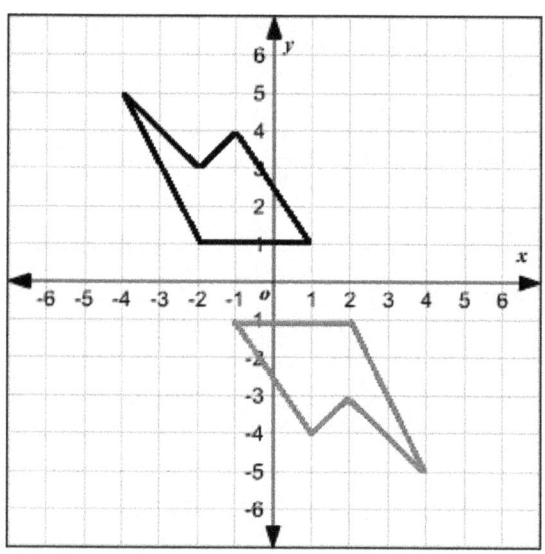

Common Core Math Practice Grade 7

3)

4)

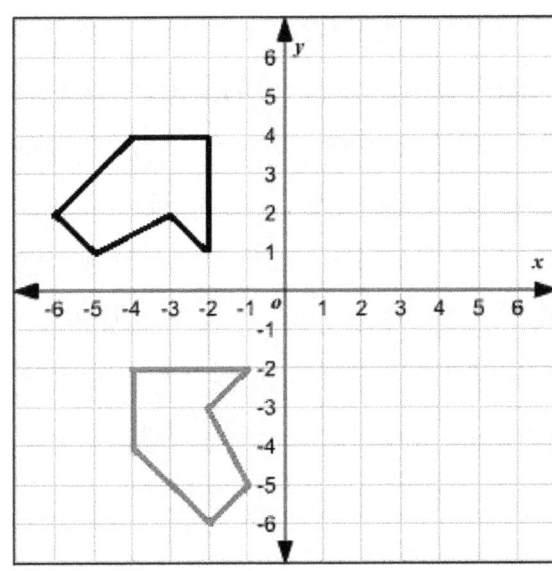

5) rotation 180° counter clockwise about the origin

6) rotation 270° about the origin

7) rotation 90° counter clockwise about the origin

8) rotation 270° counter clockwise about the origin

Dilations

1)

2)

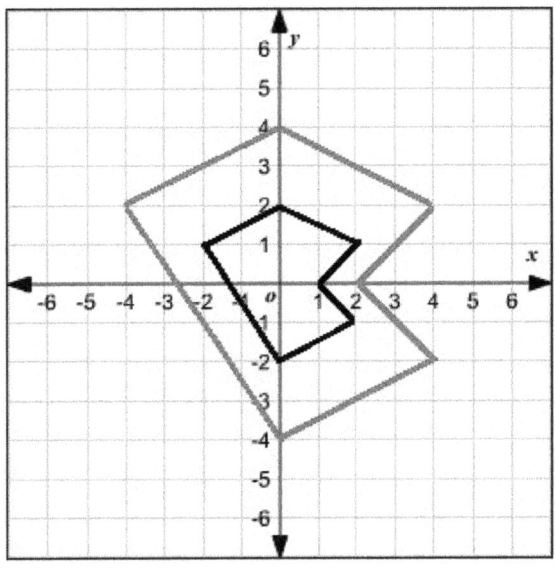

3) Reduction, $k = \frac{3}{2}$, $x = 10\ cm$

4) Enlargement, $k = \frac{1}{2}$, $x = 7.5\ cm$

Coordinate of Vertices

1) $A'(-2,-2), B'(0,-5), C'(4,-3)$

2) $D'(2,-2), E'(6,-10), F'(8,-4), G'(4,-12)$

3) $P'(-2,1), Q'(-1,4), R'(-5,5), S'(2,0)$

4) $J'(4,3), K'(2,6), L'(-3,-1)$

5) $C'(-1,4), D'(-5,2), W'(2,6), Y'(8,-3)$

6) $A'(-1,4), B'(-3,6), C'(-5,5), D'(-4,3)$

7) $K'(-2,-5), L'(-6,-6), M'(-3,-4), N'(-8,-7)$

8) $A'(2,0.5), B'(-2.5,-1), C'(-3,2)$

9) $F'(-6,0), G'(-3,6), H'(6,6)$

Chapter 10:
Geometry

Area and Perimeter of Square

Find the perimeter and area of each squares.

1)

Perimeter: .... :

Area: .... :

2)

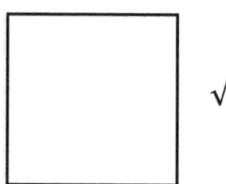

Perimeter: .... :

Area: .... :

3)

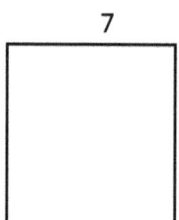

Perimeter: .... :

Area: .... :

4)

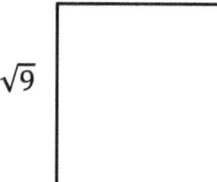

Perimeter: .... :

Area: .... :

5)

Perimeter: .... :

Area: .... :

6)

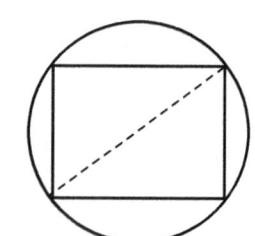

Perimeter of Square: .... :

Area of Square: .... :

Area and Perimeter of Rectangle

Find the perimeter and area of each rectangle.

1)

6
3

Perimeter: _____:
Area: _____:

2)

12
8

Perimeter: _____:
Area: _____:

3)

15
10

Perimeter: _____:
Area: _____:

4)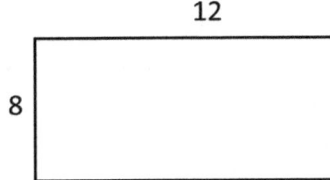

7
2.5

Perimeter: _____:
Area: _____:

5)

$1\frac{2}{5}$
$\frac{5}{7}$

Perimeter: _____:
Area: _____:

6)

5
2

Perimeter: _____:
Area: _____:

Area and Perimeter of Triangle

Find the perimeter and area of each triangle.

1)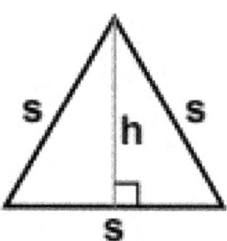

Perimeter: _____
Area: _____

2)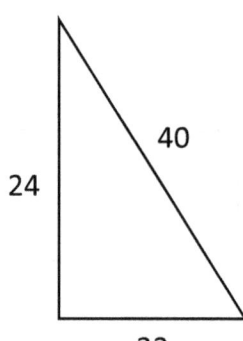

Perimeter: _____
Area: _____

3)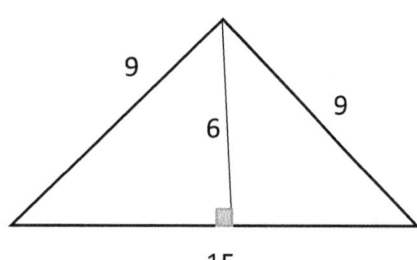

Perimeter: _____
Area _____

4) s=8

h=6

Perimeter: _____
Area: _____

5)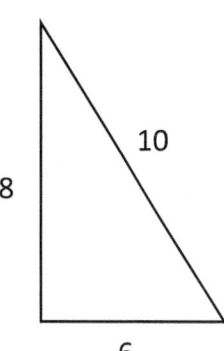

Perimeter: _____
Area: _____

6)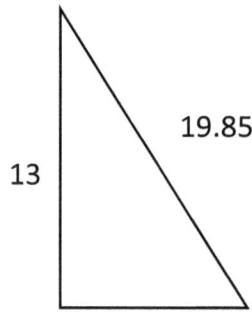

Perimeter: _____
Area: _____

Area and Perimeter of Trapezoid

Find the perimeter and area of each trapezoid.

1)

Perimeter:............:

Area:............:

2)

Perimeter:............:

Area:............:

3)

Perimeter:............:

Area ....:

4)

Perimeter:............:

Area:............:

5)

Perimeter:............:

Area:............:

6)

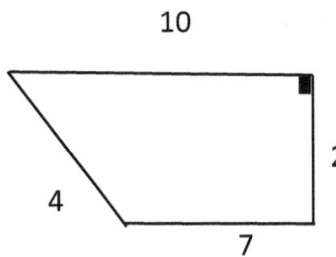

Perimeter:............:

Area:............:

Area and Perimeter of Parallelogram

Find the perimeter and area of each parallelogram.

1)

Perimeter: _____.

Area: _____.

2)

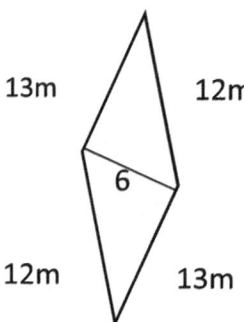

Perimeter: _____.

Area: _____.

3)

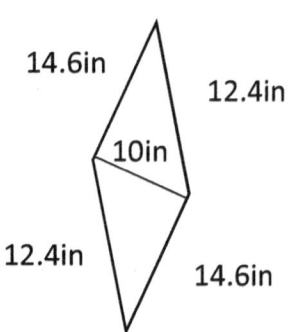

Perimeter: _____.

Area _____.

4)

Perimeter: _____.

Area: _____.

5)

24.5m

18m

Perimeter: _____.

Area: _____.

6)

Perimeter: _____.

Area: _____.

Circumference and Area of Circle

Find the circumference and area of each ($\pi = 3.14$).

1)

Circumference:

Area:

2)

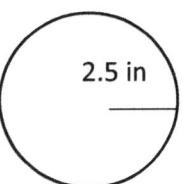

Circumference: ....

Area: ....

3)

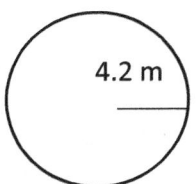

Circumference: ....

Area ....

4)

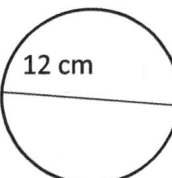

Circumference: ....

Area: ....

5)

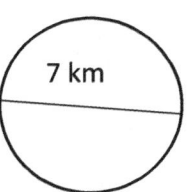

Circumference: ....

Area: ....

6)

Circumference: ....

Area: ....

Perimeter of Polygon

Find the perimeter of each polygon.

1)

Perimeter: _____ .

2)

Perimeter: _____ .

3)

Perimeter: _____ .

4)

Perimeter: _____ .

5)

Perimeter: _____ .

6)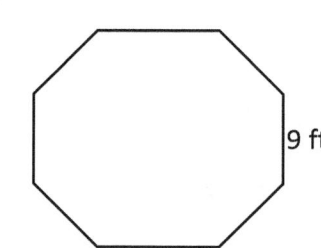

Perimeter: _____ .

Volume of Cubes

Find the volume of each cube.

1)

V:_____.

2)

V:_____.

3)

V:_____.

4)

V:_____.

5)

V:_____.

6)

V:_____.

Volume of Rectangle Prism

Find the volume of each rectangle prism

1)

V:................................:

2)

V:................................:

3)

V:................................:

4)

V:................................:

5)

V:................................:

6)
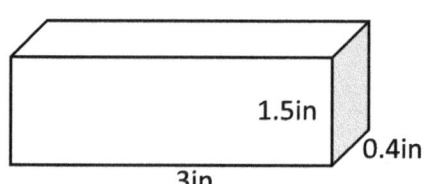

V:................................:

Common Core Math Practice Grade 7

Volume of Cylinder

Find the volume of each cylinder.

1)

V:_____:

2)

V:_____:

3)

V:_____:

4)

V:_____:

5)

V:_____:

6)
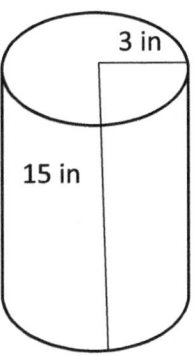

V:_____:

Volume of Spheres

Find the volume of each spheres ($\pi = 3.14$).

1)

V:..

2)

V:..

3)

V:..

4)

V:..

5)

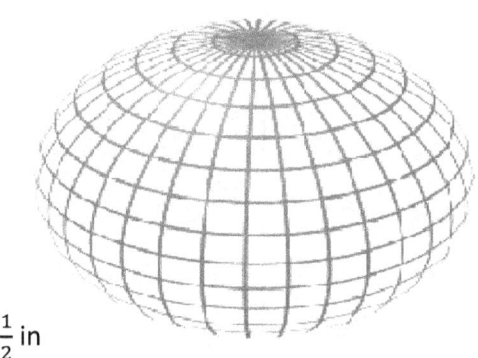

r = $3\frac{1}{2}$ in

V:..

6)

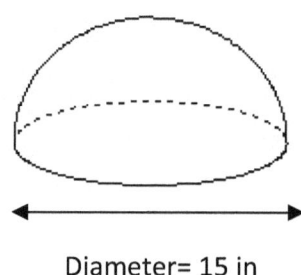

Diameter= 15 in

V:..

Common Core Math Practice Grade 7

Volume of Pyramid and Cone

Find the volume of each pyramid and cone ($\pi = 3.14$).

1)

V:..................................

2)

V:..................................

3)

V:..................................

4)

V:..................................

5)

V:..................................

6)

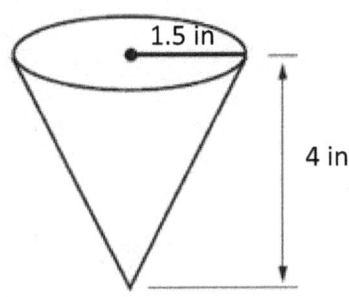

V:..................................

WWW.MathNotion.com

141

Surface Area Cubes

Find the surface area of each cube.

1)

SA:_____.

2)

SA:_____.

3)

SA:_____.

4)

SA:_____.

5)

SA:_____.

6)

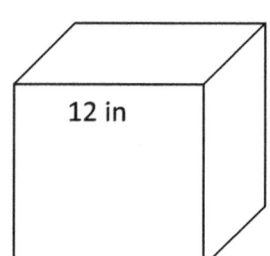

SA:_____.

Common Core Math Practice Grade 7

Surface Area Rectangle Prism

Find the surface area of each rectangular prism.

1)

SA: _____ :

2)

SA: _____ :

3)

SA: _____ :

4)

SA: _____ :

5)

SA: _____ :

6)
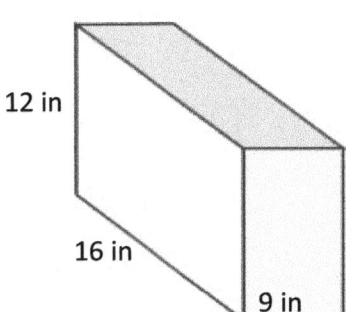

SA: _____ :

www.MathNotion.com

Surface Area Cylinder

Find the surface area of each cylinder.

1)

SA: _____.

2)

SA: _____.

3)

SA: _____.

4)

SA: _____.

5)

SA: _____.

6)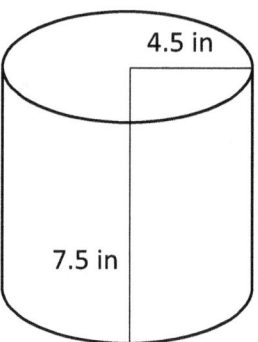

SA: _____.

Answer key Chapter 10

Area and Perimeter of Square
1. Perimeter: 20, Area: 25
2. Perimeter: $4\sqrt{6}$, Area: 6
3. Perimeter: 28, Area: 49
4. Perimeter: $4\sqrt{9}$, Area: 9
5. Perimeter: 48, Area: 144
6. Perimeter: $4\sqrt{50}$, Area: 50

Area and Perimeter of Rectangle
1- Perimeter: 18, Area: 18
2- Perimeter: 40, Area: 96
3- Perimeter: 50, Area: 150
4- Perimeter: 19, Area: 17.5
5- Perimeter: 4.23, Area: 1
6- Perimeter: 14, Area: 10

Area and Perimeter of Triangle
1- Perimeter: 3s, Area: $\frac{1}{2}sh$
2- Perimeter: 96, Area: 384
3- Perimeter: 33, Area: 45
4- Perimeter: 24, Area: 24
5- Perimeter: 24, Area: 24
6- Perimeter: 47.9, Area: 97.5

Area and Perimeter of Trapezoid
1- Perimeter: 32, Area: 49
2- Perimeter: 22, Area: 26
3- Perimeter: 44, Area: 93
4- Perimeter: 50, Area: 144
5- Perimeter: 59, Area: 245
6- Perimeter: 23, Area: 17

Area and Perimeter of Parallelogram
1- Perimeter: $26m$, Area: $20(m)^2$
2- Perimeter: $50m$, Area: $78(m)^2$
3- Perimeter: $54in$, Area: $146(in)^2$
4- Perimeter: $37cm$, Area: $63(cm)^2$
5- Perimeter: $85m$, Area: $441(m)^2$
6- Perimeter: $48m$, Area: $144(m)^2$

Circumference and Area of Circle
1) Circumference: 50.24 mm   Area: $200.96(mm)^2$
2) Circumference: 15.7in   Area: $(19.63in)^2$
3) Circumference: 26.38 m   Area: $55.39(m)^2$
4) Circumference: 37.68 cm   Area: 113.04
5) Circumference: 21.98 in   Area: $38.47(in)^2$
6) Circumference: 9.42 km   Area: $7.07(km)^2$

Perimeter of Polygon
1) 65 mm
2) 30 m
3) 57 cm
4) 59.5 in
5) 45 m
6) 72 ft

Volume of Cubes
1) $125m^3$
2) $1,000(mm)^3$
3) $216in^3$
4) $2.74(cm)^3$

WWW.MathNotion.com

5) $32.77 (ft)^3$ 6) $729 (cm)^3$

Volume of Rectangle Prism

1) $720 (cm)^3$ 3) $39.6 (m)^3$ 5) $36 (mm)^3$

2) $75 (yd)^3$ 4) $256.5 (in)^3$ 6) $1.8 (in)^3$

Volume of Cylinder

1) $339.12 (cm)^3$ 3) $1,099 (yd)^3$ 5) $635.85 (m)^3$

2) $26.69 (mm)^3$ 4) $107.94 (m)^3$ 6) $423.9 (in)^3$

Volume of Spheres

1) $1,766.25 (in)^3$ 3) $904.32 (in)^3$ 5) $179.5 (in)^3$

2) $267.95 (in)^3$ 4) $696.56 (in)^3$ 6) $883.13 (in)^3$

Volume of Pyramid and Cone

1) $512 (in)^3$ 3) $676 (in)^3$ 5) $15.19 (in)^3$

2) $1507.2 (in)^3$ 4) $301.44 (in)^3$ 6) $9.42 (in)^3$

Surface Area Cubes

1) $1,350 (in)^2$ 3) $253.5 (in)^2$ 5) $73.5 (in)^2$

2) $384 (in)^2$ 4) $144 (in)^2$ 6) $864 (in)^2$

Surface Area Rectangle Prism

1) $184 (in)^2$ 3) $312.3 (in)^2$ 5) $299.5 (in)^2$

2) $244.5 (in)^2$ 4) $942 (in)^2$ 6) $888 (in)^2$

Surface Area Cylinder

1) $169.56 (in)^2$ 3) $180.55 (in)^2$ 5) $1,657.92 (in)^2$

2) $207.24 (in)^2$ 4) $395.64 (in)^2$ 6) $339.12 (in)^2$

Chapter 11:
Statistics and probability

Mean, Median, Mode, and Range of the Given Data

Find the mean, median, mode(s), and range of the following data.

1) 26, 69, 30, 27, 19, 54, 27

Mean: __, Median: __, Mode: __, Range: __

2) 8, 12, 12, 15, 18, 20

Mean: __, Median: __, Mode: __, Range: __

3) 51, 32, 29, 33, 39, 17, 25, 29, 12

Mean: __, Median: __, Mode: __, Range: __

4) 10, 7, 3, 9, 2, 4

Mean: __, Median: __, Mode: __, Range: __

5) 20, 16, 10, 19, 13, 18, 12, 9, 9, 7

Mean: __, Median: __, Mode: __, Range: __

6) 9, 17, 18, 9, 6, 18, 8, 12

Mean: __, Median: __, Mode: __, Range: __

7) 49, 48, 86, 96, 34, 64, 48, 14, 32, 64

Mean: __, Median: __, Mode: __, Range: __

8) 45, 45, 47, 88, 89

Mean: __, Median: __, Mode: __, Range: __

9) 18, 18, 28, 36, 64

Mean: __, Median: __, Mode: __, Range: __

10) 10, 8, 2, 2, 5, 8, 1

Mean: __, Median: __, Mode: __, Range: __

11) 5, 9, 3, 5, 1, 7

Mean: __, Median: __, Mode: __, Range: __

12) 6, 7, 11, 11, 12, 12, 12

Mean: __, Median: __, Mode: __, Range: __

13) 8, 8, 0, 16, 0, 8, 16

Mean: __, Median: __, Mode: __, Range: __

14) 12, 18, 20, 7, 11, 10, 12, 16

Mean: __, Median: __, Mode: __, Range: __

15) 6, 12, 15, 15, 20

Mean: __, Median: __, Mode: __, Range: __

16) 9, 9, 12, 10, 12, 8, 17

Mean: __, Median: __, Mode: __, Range: __

17) 20, 8, 6, 9, 18, 19, 9, 6

Mean: __, Median: __, Mode: __, Range: __

18) 62, 16, 16, 28, 3, 2

Mean: __, Median: __, Mode: __, Range: __

19) 55, 22, 24, 55, 2, 4

Mean: __, Median: __, Mode: __, Range: __

20) 98, 64, 73, 86, 91, 98, 79

Mean: __, Median: __, Mode: __, Range: __

WWW.MathNotion.com

Box and Whisker Plot

1) Draw a box and whisker plot for the data set:

   24, 21, 22, 26, 24, 22, 26, 26, 30

2) The box-and-whisker plot below represents the math test scores of 20 students.

   A. What percentage of the test scores are less than 72?

   B. Which interval contains exactly 50% of the grades?

   C. What is the range of the data?

   D. What do the scores 66, 84, and 98 represent?

   E. What is the value of the lower and the upper quartile?

   F. What is the median score?

Bar Graph

Each student in class selected two games that they would like to play. Graph the given information as a bar graph and answer the questions below:

Game	Votes
Football	12
Volleyball	9
Basketball	15
Baseball	19
Tennis	15

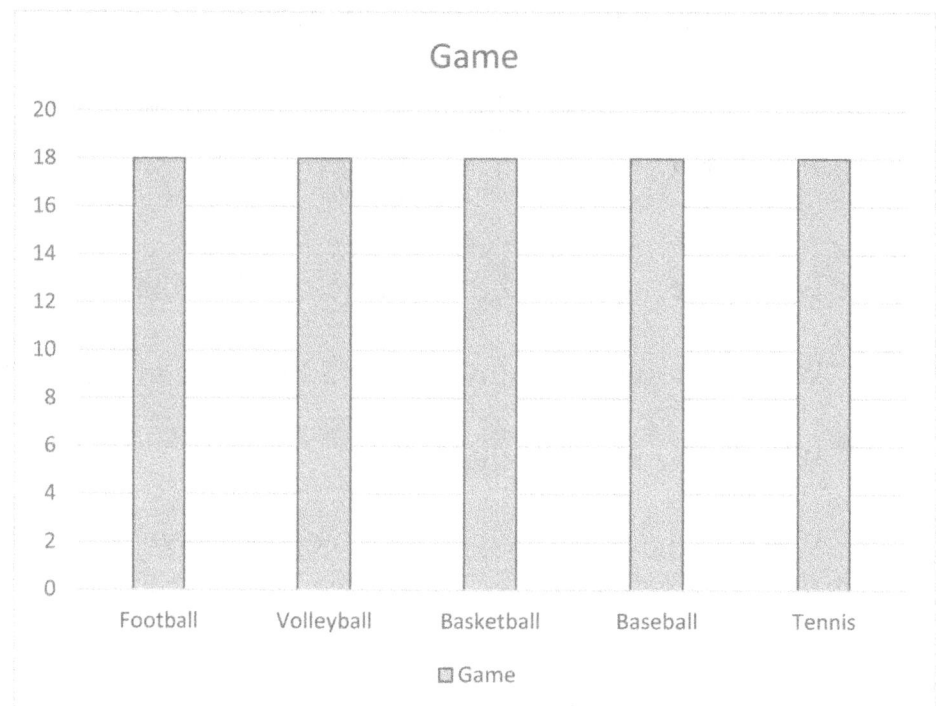

1) Which was the most popular game to play?

2) How many more student like Baseball than Football?

3) Which two game got the same number of votes?

4) How many Volleyball and Football did student vote in all?

5) Did more student like football or Tennis?

6) Which game did the fewest student like?

Dot plots

The ages of students in a Math class are given below.

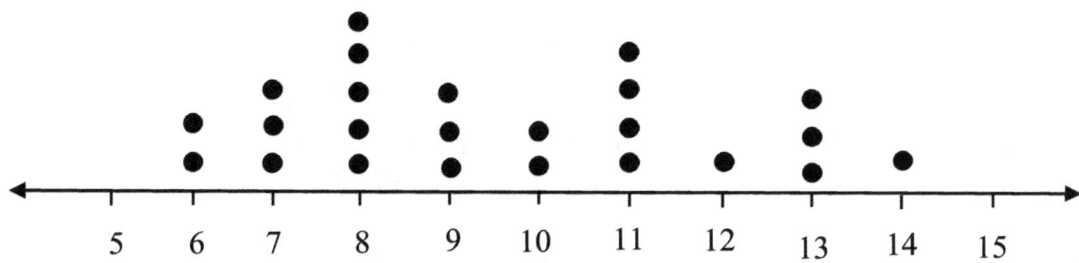

1) What is the total number of students in math class?

2) How many students are at least 11 years old?

3) Which age(s) has the most students?

4) Which age(s) has the fewest student?

5) Determine the median of the data.

6) Determine the range of the data.

7) Determine the mode of the data.

Scatter Plots

A person charges an hourly rate for his services based on the number of hours a job takes.

Hours	Rate
1	$25
2	$22.50
3	$21.50
4	$20

Hours	Rate
5	$19.50
6	$18
7	$17.50
8	$17

1) Draw a scatter plot for this data.

2) Does the data have positive or negative correlation?

3) Sketch the line that best fits the data.

4) Find the slope of the line.

5) Write the equation of the line using slope-intercept form.

6) Using your prediction equation: If a job takes 10 hours, what would be the hourly rate?

Stem–And–Leaf Plot

Make stem-and-leaf plots for the given data.

1) 22, 26, 28, 21, 42, 24, 48, 47, 29, 24, 19, 12, 45

Stem	leaf

2) 52, 54, 27, 31, 52, 24, 36, 58, 38, 34, 39, 32

Stem	leaf

3) 113, 106, 95, 95, 100, 115, 92, 114, 98, 112, 96, 107

Stem	leaf

4) 22, 15, 27, 21, 79, 24, 70, 77, 29, 24, 19, 12

Stem	leaf

5) 66, 69, 123, 67, 19, 126, 120

Stem	leaf

6) 112, 87, 96, 85, 110, 117, 92, 114, 88, 112, 98, 90

Stem	leaf

Pie Graph

80 people were survey on their favorite ice cream. The pie graph is made according to their responses. Answer following questions based on the Pie graph.

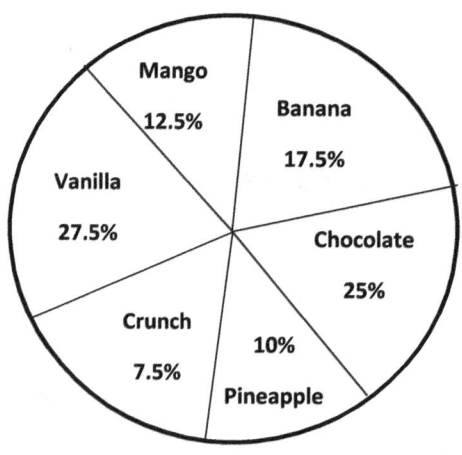

1) How many people like to eat Banana ice cream? _____

2) Approximately, which two ice creams did about half the people like the best? _____

3) How many people said either mango or crunch ice cream was their favorite? _____

4) How many people would like to have chocolate ice cream? _____

5) Which ice cream is the favorite choice of 22 people? _____

Probability

1) A jar contains 12 caramels, 7 mints and 16 dark chocolates. What is the probability of selecting a mint?

2) If you were to roll the dice one time what is the probability it will NOT land on a 2?

3) A die has sides are numbered 1 to 6. If the cube is thrown once, what is the probability of rolling a 6?

4) The sides of number cube have the numbers 3, 5, 7, 3, 5, and 7. If the cube is thrown once, what is the probability of rolling a 5?

5) Your friend asks you to think of a number from eight to twenty. What is the probability that his number will be 13?

6) A person has 5 coins in their pocket. A dime, 2 pennies, a quarter, and a nickel. If a person randomly picks one coin out of their pocket. What would the probability be that they get a penny?

7) What is the probability of drawing an odd numbered card from a standard deck of shuffled cards?

8) 24 students apply to go on a school trip. Three students are selected at random. what is the probability of selecting 3 students?

Answer key Chapter 11

Mean, Median, Mode, and Range of the Given Data

1) mean: 36, median: 27, mode: 27, range: 50
2) mean: 14.17, median: 13.5, mode: 12, range: 12
3) mean: 29.7, median: 29, mode: 29, range: 39
4) mean: 5.83, median: 5.5, mode No mode. range: 8
5) mean: 13.3, median: 12.5, mode: 9, range: 13
6) mean: 12.125, median: 10.5, mode: 9,18, range: 12
7) mean: 53.5, median: 48.5, mode: 48 and 64, range: 82
8) mean: 62.8, median: 47, mode: 45, range: 44
9) mean: 32.8, median: 28, mode: 18, range: 46
10) mean: 5.1, median: 5, mode: 2,8, range: 9
11) mean: 5, median: 5, mode: 5, range: 8
12) mean: 10.14, median: 11, mode: 12, range: 6
13) mean: 8, median: 8, mode: 8, range: 16
14) mean: 13.25, median: 12, mode: 12, range: 13
15) mean: 13.6, median: 15, mode: 15, range: 14
16) mean: 11, median: 10, mode: 9,12, range: 9
17) mean: 11.88, median: 9, mode: 6,9, range: 14
18) mean: 21.17, median: 16, mode: 16, range: 60
19) mean: 27, median: 23, mode: 55, range: 53
20) mean: 84.14, median: 86, mode: 98, range: 34

Box and Whisker Plot

1)

2)
- A. 25%
- B. 72-84
- C. 32
- D. Minimum, Median, and Maximum

E. Lower (Q_1) is 72 and upper (Q_3) is 88   F. 84

Bar Graph

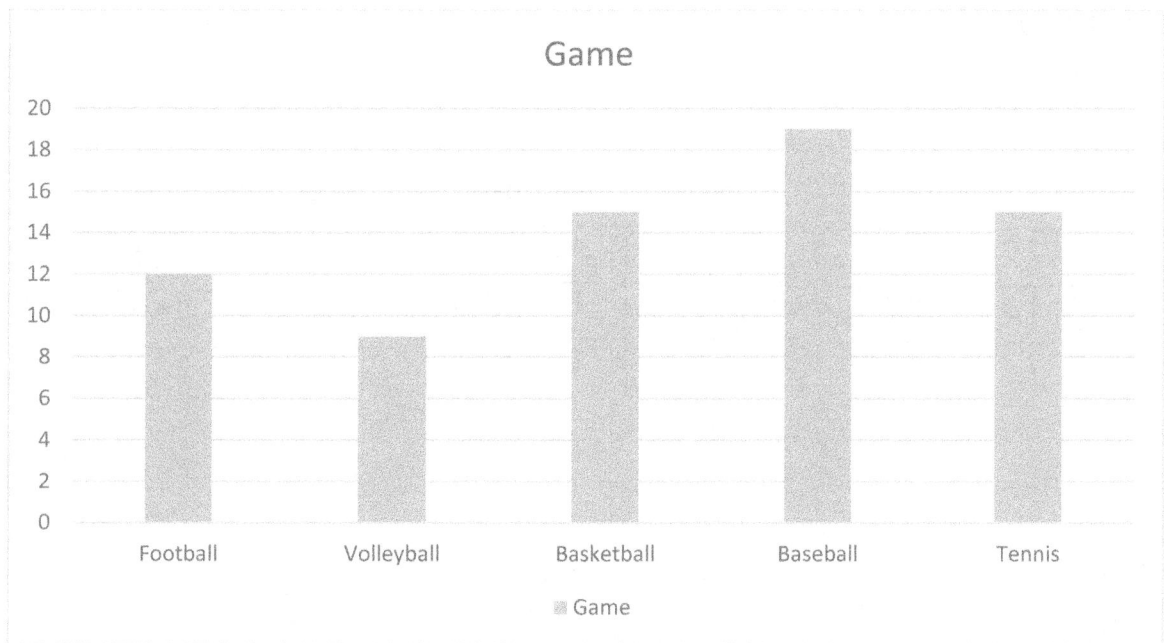

1) Baseball
2) 7 students
3) Basketball and Tennis
4) 21
5) Tennis
6) Volleyball

Dot plots

1) 24
2) 9
3) 8
4) 12 and 14
5) 3
6) 8
7) 3

Scatter Plots

1)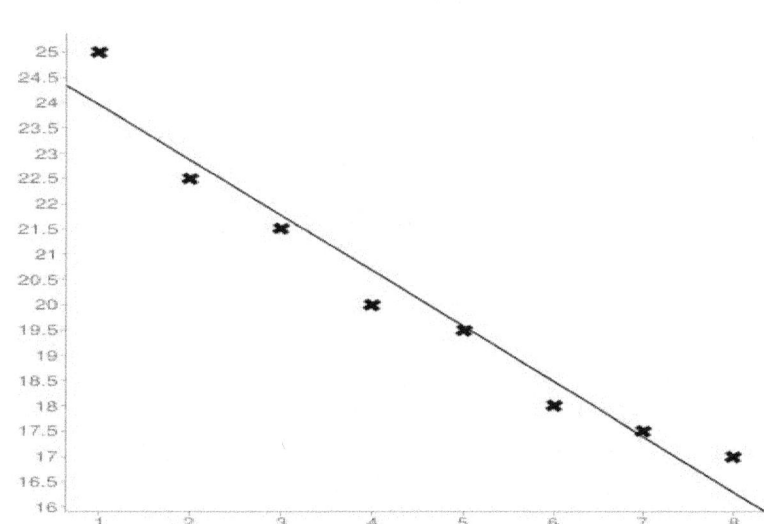

Common Core Math Practice Grade 7

2) Negative correlation

3) ----

4) Slope(m)= -1

5) $y = -x + 25$

6) 15

Stem–And–Leaf Plot

1)

Stem	leaf
1	2 9
2	1 2 4 4 6 8 9
4	2 7 8 5

2)

Stem	leaf
2	4 7
3	1 2 4 6 8 9
5	2 2 4 8

3)

Stem	leaf
9	2 5 5 6 8
10	0 6 7
11	2 3 4 5

4)

Stem	leaf
1	2 9 5
2	1 2 4 4 7 9
7	0 7 9

5)

Stem	leaf
1	9
6	6 7 9
12	0 3 6

6)

Stem	leaf
8	5 7 8
9	0 2 6 8
11	0 2 2 4 7

Pie Graph

1) 14

2) Vanilla and chocolate

3) 16

4) 20

5) Vanilla

Probability

1) $\frac{1}{5}$

2) $\frac{5}{6}$

3) $\frac{1}{6}$

4) $\frac{1}{3}$

5) $\frac{1}{12}$

6) $\frac{2}{5}$

7) $\frac{4}{13}$

8) $\frac{1}{8}$

Common Core Test Review

Common Core GRADE 7 MAHEMATICS REFRENCE MATERIALS

Linear Equations

Slope-intercept form $\qquad y = mx + b$

Constant of proportionality $\qquad k = \frac{y}{x}$

Circumference

Circle $\qquad C = 2\pi r \quad$ or $\quad C = \pi d$

Area

Triangle $\qquad A = \frac{1}{2}bh$

Rectangle or Parallelogram $\qquad A = bh$

Trapezoid $\qquad A = \frac{1}{2}h(b_1 + b_2)$

Circle $\qquad A = \pi r^2$

Volume

Prism or cylinder $\qquad V = Bh$

Pyramid or Cone $\qquad V = \frac{1}{3}Bh$

Additional Information

Pi $\qquad \pi = 3.14 \quad$ or $\quad \pi = \frac{22}{7}$

Distance $\qquad d = rt$

Simple interest $\qquad I = prt$

Compound interest $\qquad I = p(1 + r)^t$

Common Core GRADE 7 MAHEMATICS REFRENCE MATERIALS

LENGTH

Customary

1 mile (mi) = 1,760 yards (yd)

1 yard (yd) = 3 feet (ft)

1 foot (ft) = 12 inches (in.)

Metric

1 kilometer (km) = 1,000 meters (m)

1 meter (m) = 100 centimeters (cm)

1 centimeter(cm) = 10 millimeters(mm)

VOLUME AND CAPACITY

Customary

1 gallon (gal) = 4 quarts (qt)

1 quart (qt) = 2 pints (pt.)

1 pint (pt.) = 2 cups (c)

1 cup (c) = 8 fluid ounces (Fl oz)

Metric

1 liter (L) = 1,000 milliliters (mL)

WEIGHT AND MASS

Customary

1 ton (T) = 2,000 pounds (lb.)

1 pound (lb.) = 16 ounces (oz)

Metric

1 kilogram (kg) = 1,000 grams (g)

1 gram (g) = 1,000 milligrams (mg)

Common Core Practice Test 1

Mathematics

GRADE 7

Administered *Month Year*

Common Core Math Practice Grade 7

1) Peter paid for 7 sandwiches.

   - Each sandwich cost 10.45.

   - He paid for 6 bags of fries that each cost $1.23

   Which equation can be used to determine the total amount, y, Peter paid?

   A. $y = 7(10.45) + 6(1.23)x$

   B. $y = (10.45 + 1.23)x$

   C. $y = 7(10.45) + 6(1.23)$

   D. $y = 10.45x + 6(1.23)$

2) What is the decimal equivalent of the fraction $\frac{40}{54}$?

   A. 0.74

   B. $0.7\overline{40}$

   C. $0.\overline{740}$

   D. 0.740

3) The circumference of a circle is 12π centimeters. What is the area of the circle in terms of π?

   A. 25π

   B. 36π

   C. 144π

   D. 49π

Common Core Math Practice Grade 7

4) What is the volume of rectangular prism when the two triangular prisms below are stuck together?

   A. 504 in^3

   B. 405 in^3

   C. 544 in^3

   D. 50.4 in^3

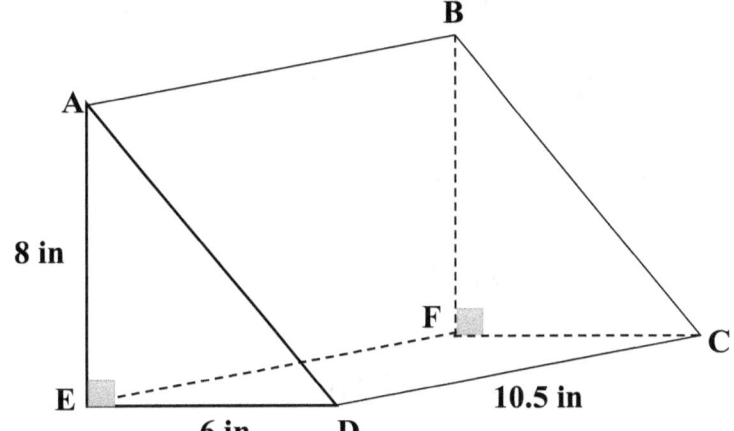

5) Which number line Shows the solution to the inequality $-5x + 4 < -6$?

   A.

   B.

   C.

   D.

6) What is the value of $(2 + 6)^2 + (2 - 6)^2$?

   A. −64

   B. 16

   C. 128

   D. 80

Common Core Math Practice Grade 7

7) Arsan has $9 to spend on school supplies. The following table shows the price of each item in the school store. No sale tax is charged on these items. Which the combination of items can Arsan buy with his $9?

A. 4 Notebooks and 2 Pens

B. 4 Folders and 5 Erasers

C. 6 Notebooks and 2 Folders

D. 5 Erasers and 7 Pens.

Item	Price
Notebook	$2.20
Pen	$0.60
Eraser	$ 0.45
Folder	$1.86

8) If 12% of x is 72, what is 45% of x?

   A. 270

   B. 27.45

   C. 17.12

   D. 2.745

9) If all variables are positive, find the square root of $\frac{16x^{10}y^6}{36x^2y^2}$?

   A. $\frac{4}{9}x^4y^2$

   B. $\frac{4y^2}{3x^4}$

   C. $\frac{2}{3}x^4y^2$

   D. $3\frac{2}{3}x^2y^4$

WWW.MathNotion.com

Common Core Math Practice Grade 7

10) Which is closest to the perimeter of the right triangle in the figure below?

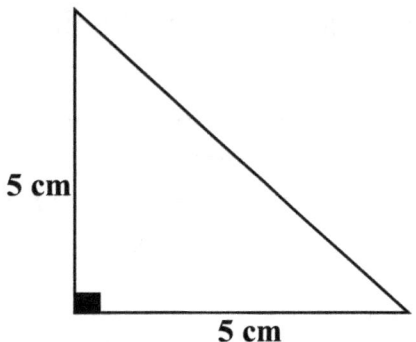

A. 19.4

B. 17.4

C. 19

D. 17

11) What is the range of the following set of data: 2, −2, 10, −4, 6, 12, 10, 4, 16?

A. 10

B. 12

C. 24

D. 20

12) Alex starts a saving account with $60. Every week he adds $7 to his account. Which equation can be used to determine the number of weeks w, after which Alex's accounts reaches $198?

A. $7w + 198 = 60$

B. $7 + w = 198$

C. $7w + 60 = 198$

D. $7w - 60 = 198$

13) The medals won by United States, Australia and Spain during a basketball competition are shown in the table below:

United States	Australia	Spain
10	11	4

Out of the medals won by these three countries, what percentage of medals did the United States win?

A. 30%

B. 40%

C. 45%

D. 90%

14) A school has 760 students and 42 chemistry teachers and 19 physics teachers. What is the ratio between the number of physics teachers and the number of students of the school?

A. $\frac{1}{40}$

B. $\frac{2}{17}$

C. $\frac{1}{18}$

D. $\frac{1}{19}$

15) James has his own lawn mowing service. The maximum James charges to mow a lawn is $45. Which inequality represents the amount James could charge, P, to mow a lawn?

　A. P ≥ 45

　B. P ≤ 45

　C. P = 45

　D. P < 45

16) The attendance at three clubs was 861, 446, and 677. Which expression shows how to estimate the total attendance at the clubs?

　A. 800 + 400 + 600

　B. 800 + 500 + 600

　C. 900 + 400 + 700

　D. 900 + 400 + 600

17) A diver started its dive at sea level and descended 42 feet in half hour. Which integer represents the diver's depth after 55 minutes?

　A. −42

　B. −77

　C. 21

　D. 77

18) What is 426,000,000 in scientific notations?

   A. 426×10^6

   B. 42.6×10^7

   C. 4.26×10^8

   D. 0.426×10^9

19) The ratio of boys to girls in Maria Club is the same as the ratio of boys to girls in Hudson Club. There are 18 boys and 54 girls in Maria Club. There are 25 boys in Hudson Club. How many girls are in Hudson Club?

   A. 20

   B. 18

   C. 25

   D. 75

20) A girl in State A spent $56 before a 4.25% sales tax and a girl in State B spent $32 before an 6.75% sales tax. How much more money did the girl from State A spend than the girl from State B after sales tax was applied? Round to the nearest hundredth.

   A. 24.22

   B. 22.24

   C. 20.25

   D. 18.24

21) On average, Simone drinks $\frac{2}{9}$ of a 18-ounce glass of coffee in $\frac{1}{4}$ hour. How much coffee does she drink in an hour?

   A. 25 ounces

   B. 19 ounces

   C. 16 ounces

   D. 20 ounces

22) Line P, R, and S intersect each other, as shown in below diagram. Based on the angle measures, what is the value of θ?

   A. 155°

   B. 133°

   C. 121°

   D. 106°

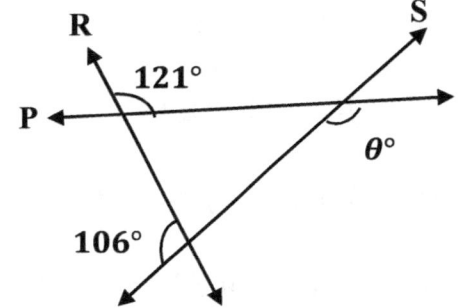

23) Which expression is represented by the model below?

   A. $-4 \cdot (-9)$

   B. $(-4) \cdot 9$

   C. $4 \cdot (-9)$

   D. $4 \cdot 9$

Common Core Math Practice Grade 7

24) A circular table has the circumference of 26π cm. What is the area of the table?

   A. 169π

   B. 144π

   C. 26π

   D. 676π

25) Lucas walks down 6 flights of stairs. Each flight has 14 steps. Which describes Lucas's descent?

   A. $(-6) \times 14 = -84$

   B. $(-6) + 14 = 8$

   C. $(-6)(-14) = 84$

   D. $(-6) + (-14) = -20$

26) Mia has a loan of $45,364. This loan has a simple interest rate of 4.5% per year. What is the amount of interest that Mia will be charged on this loan at the end of one year?

   A. $47,405.38

   B. $3,418

   C. $20,413

   D. $2,041.38

27) Kaylee uses 8 grams of organic juice powder to make 33 fluid ounces of juice. Last week she made 297 fluid ounces of juice. How many grams of powder did Kaylee use to make juice last week?

A. 7.5 g

B. 264 g

C. 72 g

D. 36 g

28) The table below shows the distance, y, a lion can travel in mile in x hour.

Time (x, hour)	Distance (y, mile)
5	260
10	520
15	780
20	1,040
25	1,300

Based on the information in the table, which equation can be used to model the relationship between x and y?

A. $y = x + 5$

B. $y = 5x$

C. $y = x + 260$

D. $y = 52x$

29) An automobile designer built a scale model of a car using a scale in which 4 mm represents 0.32 m. The height of the car is 1.28 m. What is the height of the scale model in millimeter?

A. 32 mm

B. 5 mm

C. 16 mm

D. 72 mm

30) Which diagram best represents the relationship among integers, rational numbers, whole numbers and -151?

A.

C.

B.

D.

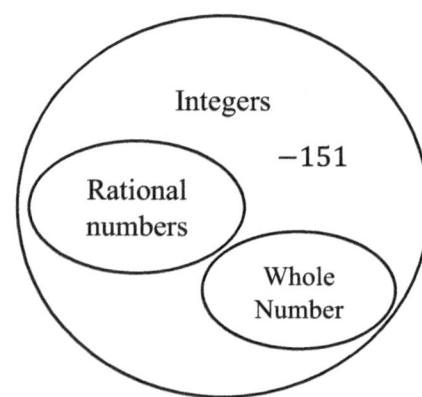

31) The spinner shown has eight congruent sections.

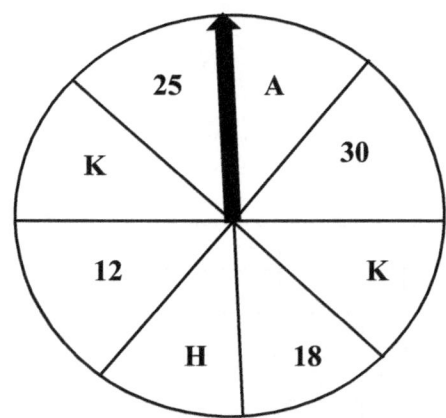

The spinner is spun 180 times. What is a reasonable prediction for the number of times the spinner will land on a letter?

A. 45

B. 18

C. 120

D. 90

32) Linda has $12\frac{3}{4}$ bag of flour to bake bread. Each bag will cover $10.22\ cm^2$ breads. How many cm^2 breads will Linda be able to bake if she uses all these cups of flour?

A. $130.305\ cm^2$

B. $130.035 cm^2$

C. $150.035 cm^2$

D. $150.305 cm^2$

33) Which graph best represents the distance a car travels when going 25 miles per hour?

A.

B.

C.

D.

34) The temperature is shown in the table below, on each of day in the week for a city in February. What is the mean temperature, in the city for that week?

A. −14

B. −14.2

C. −2

D. −5.4

Day	Temperature (°F)
Monday	−23
Tuesday	−27
Wednesday	−19
Thursday	3
Friday	18
Saturday	−3
Sunday	37

35) What is the value of this expression $46 \div 0.92$?

A. 0.2

B. 0.5

C. 2

D. 50

36) Which arithmetic sequence is represented by the expression $5m - 2$, where m represents the position of a term in the sequence?

A. 10, 15, 20, 25, 30, …

B. 13, 18, 23, 28, 33, …

C. 8, 13, 17, 20, 25, …

D. 3, 8, 12, 15, 18, …

37) Which equation is true when $d = -22$?

   A. $3d - 20 = -35$

   B. $-25 + 2d = 18$

   C. $\frac{d}{11} + 12 = 23$

   D. $\frac{d}{2} + 35 = 24$

38) The dot plots show how many minutes per day do 7th grade study math after school at two different schools on one day.

   Number of minuets study in school 1

   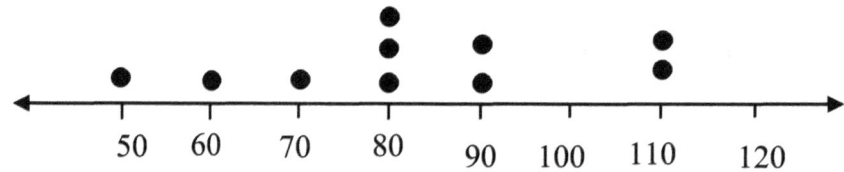

   Number of minuets study in school 2

   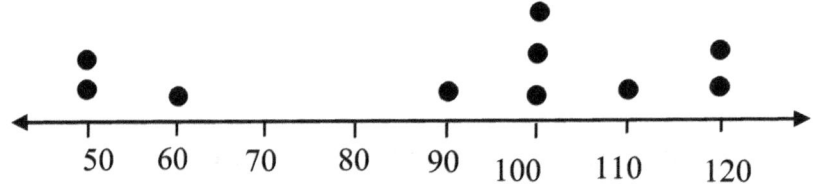

   Which statement is supported by the information in the dot plots?

   A. The mode of the data for School 2 is greater than the mode of the data for School 1.

   B. The mean of the data for School 1 is greater than the mean of the data for School 2.

   C. The median of the data for School 2 is smaller than the median of the data for School 1.

   D. The median and mean of the data for two schools are equal.

39) Which expression is equivalent to $-40 - 120d$?

   A. $-10(3 - 15d)$

   B. $-4(40d - 8)$

   C. $-8(5 + 15d)$

   D. $-160\ d$

40) Which number represents the probability of an event that is very likely to occur?

   A. 0.26

   B. 1.2

   C. 0.73

   D. 0.03

Common Core Practice Test 2

Mathematics

GRADE 7

Administered *Month Year*

Common Core Math Practice Grade 7

1) What is the decimal equivalent of the fraction $\frac{29}{22}$?

   A. 1.18

   B. $1.\overline{318}$

   C. $1.3\overline{18}$

   D. 1.318

2) Thomas is shareholder of a company. The price of stock is $89.76 on the morning of day 1. Thomas records the change in the price of the stock in the chart below at the end of each day, but some information is missing.

Day	Change in Price ($)
1	+ 0.92
2	+0.86
3	
4	−0.84
5	

   The change in the price for day 3 is $\frac{3}{4}$ of the change in the price for day 4. At the end of day 5, the price of Thomas's stock is $91.03. What is the change, in dollars, in the price of the stock for day 5?

   A. −0.98

   B. 0.98

   C. 0.05

   D. 1.05

3) Kevin adds $\frac{3}{4}$ cups of sugar into a mixture every $\frac{1}{5}$ hour. What is the rate, in cups per minute, at which Kevin adds sugar to the mixture?

A. $\frac{1}{16}$

B. $2\frac{1}{4}$

C. $\frac{1}{48}$

D. $\frac{1}{12}$

4) A box of ball contains 7 blue balls, 5 red balls, 10 black balls, and 2 green balls. All the balls are the same size and shape. Brian will select a ball at random. Which of the following best describes the probability that Brian will select a green ball?

A. impossible

B. certain

C. unlikely

D. Likely

5) The first number in a pattern is 6. Each following number is found by subtracting 9 from the previous number. What is the fifth number in the pattern?

A. −39

B. −21

C. −19

D. −30

Common Core Math Practice Grade 7

6) What is the value of this expression?

$$13 \div 0.5$$

A. 8

B. 26

C. 0.06

D. 0.26

7) A car dealer purchased a car 5 years ago for $28,000. Now, the value of the car has decreased by 75%. What is the value of the car now?

A. $11,000

B. $7,700

C. $7,000

D. $28,000

8) Mr. Turner is digging a trench to put in the new school sprinkler system. Every $\frac{1}{6}$ hour, the length of his trench increases by $\frac{3}{4}$ foot. By how much does the length, in feet, of Mr. Turner's trench increase each hour?

A. $\frac{1}{9}$

B. $\frac{2}{9}$

C. $\frac{4}{18}$

D. $\frac{9}{2}$

9) Evelyn opened a bank account. She adds the same amount of money to her account each month. The table below shows the amounts of money in her account at the ends of certain numbers of months.

How much money does Evelyn add to her bank account each month?

A. $12

B. $9

C. $18

D. $36

Month	Amount
5	$90
9	$162
13	$234

10) Brendan charges $42 per hour plus $70 to enter data. He accepted a project for no more than $680. Which inequality can be used to determine all the possible numbers of hours (x) it took the man to enter the data?

A. $42x + 70 \leq 680$

B. $42x + 70 > 680$

C. $70x + 42 < 680$

D. $70x + 42 \geq 680$

11) Simplify: $7x - 4.3y - 8x - x + 4y$

A. $-3.2xy$

B. $2.8x - 3y$

C. $-2x + 0.3y$

D. $-2x - 0.3y$

12) Use the coordinate grid below to answer the question. What is the circumference of the circle?

A. 125.6

B. 12.56

C. 25.12

D. 25.6

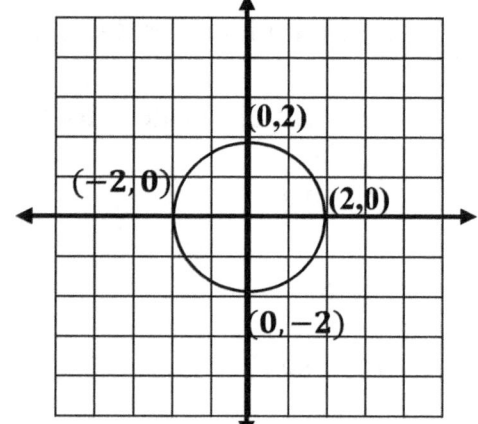

13) Robert sold botanical juice bottles in packages of 7. On the first day, he sold x packages. On the next day, he sold triple as many packages as on the first day. Which expression shows the number of juice bottles he sold on the second day?

A. $\frac{7x}{3}$

B. $7(3x)$

C. $7(x+3)$

D. $7(x-3)$

14) What is the value of the expression $-3x + 4$ when $x = 5$?

A. 11

B. -11

C. -19

D. 19

15) Using data from house sales, probabilities for the story of a house sold were calculated. The probabilities for two story are listed below.

- The probability a house sold has one story is 0.36.
- The probability a house sold has two story 0.45.

Based on these probabilities, how many of the next 200 houses sold are likely to be one story and how many are likely to be two story?

A. one story: 36, two story: 45

B. one story: 114, two story: 45

C. one story: 72, two story: 90

D. one story: 90, two story: 200

16) Ms. Wilson drove from her workplace to her mother's house. The distance (y), in miles, she drove based on the time (x), in hours, is graphed on a coordinate grid. Her drive is represented by the line segment with endpoints at (0, 0) and $(2\frac{2}{5}, 144)$. Based on the point on the graph with an x-coordinate of 1, which statement must be true?

A. Ms. Wilson drove $\frac{1}{60}$ of the distance from her house to her sister's house

B. Ms. Wilson drove at a unit rate of 60 miles per hour.

C. Ms. Wilson drove 1 mile in 60% of an hour.

D. Ms. Wilson drove from her house to her mother's house in 1 hour.

17) The temperature is 12° F. As a cold front move in, the temperature drops 5° F per half hour. What is the temperature at the end of 3 hours?

A. 12°F

B. 18°F

C. −20°F

D. −18°F

18) A printer originally cost h dollars, including tax. Eddy purchased the printer when it was on sale for 46% off its original cost. Which of the following expressions represents the final cost, in dollars, of the printer Eddy purchased?

A. $h + 0.54$

B. $h - 0.46$

C. $0.54h$

D. $0.46h$

19) Use the set of data below. What is the median of the list of numbers?

$$37, 22, 34, 27, 22, 29$$

A. 34

B. 28

C. 27

D. 22

20) Of the 160 people in a group, 30% eat organic food. What is the total number of people eat non-organic food in the group?

   A. 135

   B. 70

   C. 160

   D. 112

21) Asher worked out at a gym for 2 hours. His workout consisted of jogging for 32 minutes, playing volleyball for 58 minutes, and playing billiards for the remaining amount of time. What percentage of Asher's workout was spent playing billiards?

   A. 25%

   B. 30%

   C. 90%

   D. 55%

22) Multiply: $3\frac{3}{13} \times \frac{-5}{13}$

   A. $3\frac{15}{13}$

   B. $-11\frac{5}{13}$

   C. $-\frac{220}{169}$

   D. 6

23) Triangle PRS is shown on the grid below

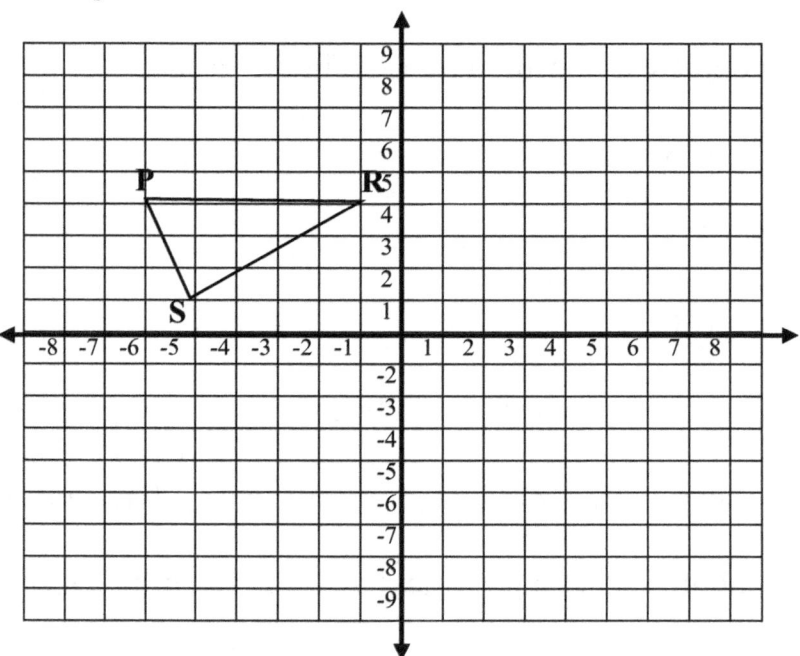

If triangle PRS is reflected across the y-axis to form triangle P′R′S′, which ordered pair represents the coordinates of R′?

A. $(-1, -4)$

B. $(1, -4)$

C. $(1, 4)$

D. $(4, -1)$

24) Benjamin eats $\frac{3}{8}$ of a cake. What percent of the cake does he eat?

A. 37.5%

B. 37%

C. 3.80%

D. 38.5%

Common Core Math Practice Grade 7

25) The angle measures of a triangle GBD are shown in the diagram. What is the value of ∠B?

   A. 34.5

   B. 66

   C. 24

   D. 64

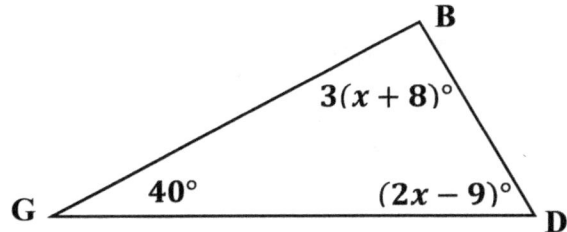

26) In a party people drink 48.20 liters of juices. There are approximately 29.57 milliliters in 1 fluid ounce. Which measurement is closest to the number of fluid ounces in 59.15 liters?

   A. 0.163 fl oz

   B. 1,063.30 fl oz

   C. 1,06.30 fl oz

   D. 1,630 fl oz

27) Gummy candy cost $0.60 per ounce. Mary buys $3\frac{1}{2}$ ounces of gummy candy for himself and 3 ounces for her friend. What is the total cost of the gummy candy Mary buys?

   A. $2.90

   B. $3.9

   C. $3.10

   D. $6.20

Common Core Math Practice Grade 7

28) The dimensions of a square pyramid are shown in the diagram. What is the volume of the square pyramid in cubic inches?

   A. 396.75 in^3

   B. 369.25 in^3

   C. 396.25 in^3

   D. 639.75 in^3

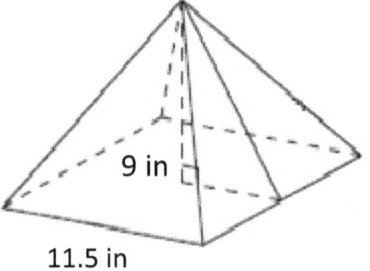

29) The store manager spent $16,230 to buy a new freezer and 32 tables. The total purchase is represented by this equation, where v stands for the value of each table purchased: $32v + 2,150 = 16,230$

   What was the cost of each table that the manager purchased?

   A. $504

   B. $340

   C. $540

   D. $440

30) In a city, at 1:30 A.M., the temperature was $-8°F$. At 1:30 P.M., the temperature was $15°F$. Which expression represents the increase in temperature?

   A. $-8 - 15$

   B. $|-8 - 15|$

   C. $|-8| - 15$

   D. $-8 - |15|$

31) A scale drawing of triangle DEF that will be used on a wall is shown below. What is the perimeter, in meter, of the actual triangle used on the wall?

Scale: 1 cm: $3\frac{1}{4}$ m

A. $48\frac{3}{4}$

B. $45\frac{3}{4}$

C. $45\frac{1}{2}$

D. $49\frac{1}{4}$

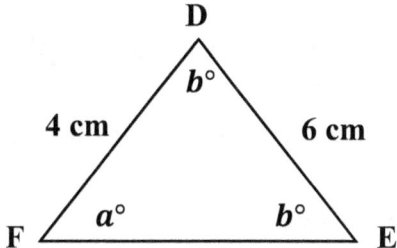

32) Water is poured to fill a pool in the shape of a rectangular prism. The pool is 15 feet long, 5.6 feet wide, and 10.75 feet high. How much cubic feet of water are needed to fill the pool?

A. $903\ ft^3$

B. $390\ ft^3$

C. $930\ ft^3$

D. $9.03\ ft^3$

33) What is the solution set for the inequality $-6x + 10 > -20$?

A. $x > 5$

B. $x < 5$

C. $x > -5$

D. $x < -5$

34) A group of employees have their weight recorded to make a data set. The mean, median, mode, and range of the data set are recorded. Then, the weight of the manager is included to make a new data set. The manager's weight is more than all but one of the employees. Which measure must be the same when the manager's weight included?

A. Mean

B. Median

C. Mode

D. Range

35) Angles α and β are complementary angles. Angles α and γ are supplementary angles. The degree measure of angle β is 110°. What is the measure of angle γ?

A. 20°

B. 35°

C. 70°

D. 90°

36) The ratio of boys to girls in Geometry class is 3 to 4. There are 20 girls in the class. What is the total number of students in Geometry class?

A. 12

B. 30

C. 45

D. 27

37) The bar graph shows a company's income and expenses over the last 5 years.

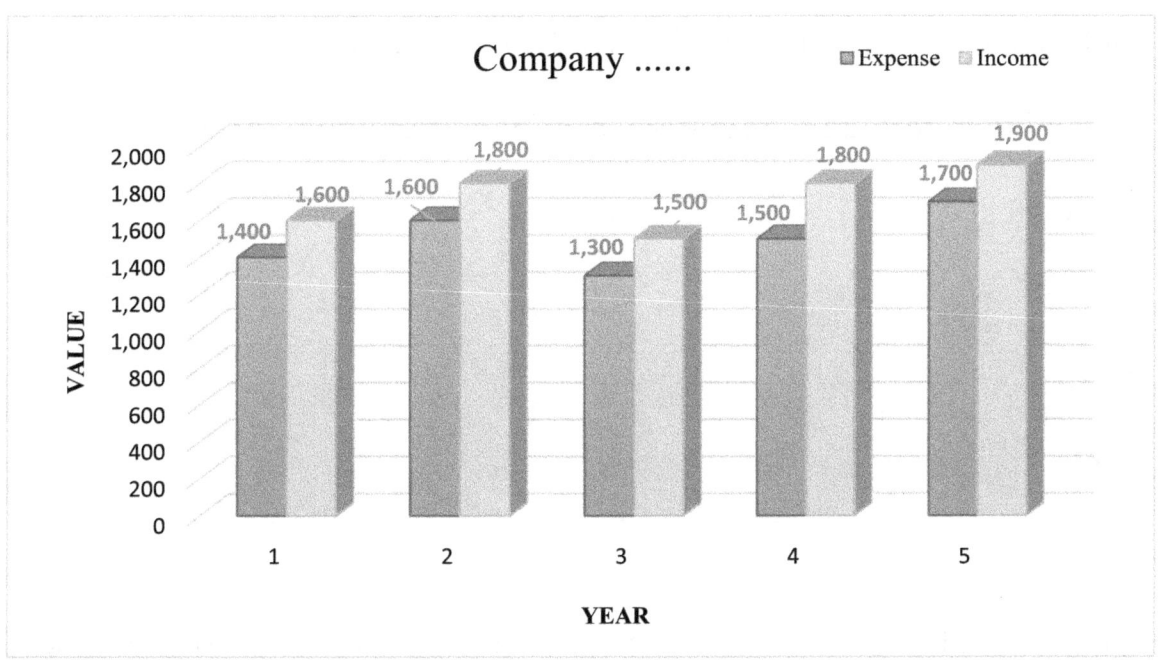

Which statement is supported by the information in the graph?

A. Expenses have increased $300 each year over the last 5 years.

B. The income in Year 5 was 20 % more than the income in Year 1.

C. The combined income in Years 3, 4, and 5 was equal to the combined expenses in Years 2, 3, and 4.

D. Expenses in the year 3 was more than half of the income in the year 4.

38) Which expression is equivalent to the $(3n - 9) - \frac{1}{3}(8 - 6n) + \frac{5}{3}$?

A. -5

B. $-5n - 10$

C. $5n - 10$

D. $5n - 8$

39) A box contains colored marbles. 6 marbles are red, 3 marbles are green, and 6 marbles are blue. A marble will be randomly selected from the box. What is the probability in decimal form that the marble selected will be green? (round nearest hundredths)

A. 0. 2

B. 0.4

C. 0.5

D. 0.8

40) Patricia bought a bottle of 16-ounce balsamic vinegar for $13.06. She used 35% of the balsamic vinegar in two weeks. Which of the following is closest to the cost of the balsamic she used?

A. $4.37

B. $2.57

C. $4.57

D. $2.37

Answers and Explanations

Common Core Math Practice Grade 7

Answer Key

Now, it's time to review your results to see where you went wrong and what areas you need to improve!

Common Core Math Practice Tests

Practice Test 1

#	Ans	#	Ans	#	Ans
1	C	16	C	31	D
2	C	17	B	32	A
3	B	18	C	33	B
4	A	19	D	34	C
5	B	20	A	35	D
6	D	21	C	36	B
7	D	22	B	37	D
8	A	23	C	38	A
9	C	24	A	39	C
10	D	25	A	40	C
11	D	26	D		
12	C	27	C		
13	B	28	D		
14	A	29	C		
15	D	30	B		

Practice Test 2

#	Ans	#	Ans	#	Ans
1	C	16	B	31	C
2	B	17	D	32	A
3	A	18	C	33	B
4	C	19	B	34	D
5	D	20	D	35	A
6	B	21	A	36	B
7	C	22	C	37	D
8	D	23	C	38	C
9	C	24	A	39	A
10	A	25	B	40	C
11	D	26	D		
12	B	27	B		
13	B	28	A		
14	B	29	D		
15	C	30	B		

Common Core Math Practice Grade 7

Practice Test 1

Answers and Explanations

1) Answer: C

Let y be the total amount paid.

We have been given that Peter bought 7 sandwiches that each cost the $10.45.

So, the cost of 7 sandwiches would be 7(10.45). He paid for 6 bags of fries that each cost the $1.23. So, the cost of 7 bags would be 8(1.23)

Then, the total cost of sandwiches and fries would be $y = 7(10.45) + 6(1.23)$

2) Answer: C

To get the answer to 40 over 54 as a decimal, we divide 40 by 54.

$\frac{40}{54} = 0.740740740 \ldots = 0.\overline{740}$

3) Answer: B

Use the formula of circumference of circles.

Circumference = πd = 2π (r) = 12 π ⇒r = 6

Radius of the circle is 6. Now, use the areas formula:

Area = πr^2 ⇒ Area = π(6)² ⇒ Area = 36 π

4) Answer: A

The volume of a triangular prism is the base times the height. $V = Bh$

Area of the base = $\frac{1}{2} b. h \rightarrow B = \frac{1}{2} \times 6 \times 8 = 24$

$V = B.h = 24 \times 10.5 = 252$ and we need two triangular prisms, then

$2 \times 252 = 504$

5) Answer: B

$-5x + 4 < -6$, add -4 to both sides $-5x < -10$ divide each term by -5

If an inequality is multiplied or divided by a negative number, you must change the direction of the inequality, then $x > 2$

WWW.MathNotion.com

Common Core Math Practice Grade 7

6) Answer: D

$(2+6)^2 + (2-6)^2 = (8)^2 + (-4)^2 = 64 + 16 = 80$

7) Answer: D

A. $(4 \times 2.20) + (2 \times 0.60) = 8.80 + 1.20 = 10 > 9$

B. $(4 \times 1.86) + (5 \times 0.45) = 7.44 + 2.25 = 9.69 > 9$

C. $(6 \times 2.20) + (2 \times 1.86) = 13.2 + 3.72 = 16.92 > 9$

D. $(5 \times 0.45) + (7 \times 0.60) = 2.25 + 4.2 = 6.45 < 9$

8) Answer: A

$0.12 \times x = 72 \rightarrow x = \frac{72}{0.12} = \frac{7,200}{12} = 600$

$45\% \, of \, 600 = 0.45 \times 600 = 270$

9) Answer: C

$\sqrt{\frac{16x^{10}y^6}{36x^2y^2}} = \sqrt{\frac{16}{36} \times \frac{x^{10}y^6}{x^2y^2}} = \sqrt{\frac{4}{9}x^8y^4} = \frac{2}{3}x^4y^2$

10) Answer: D

Use Pythagorean Theorem: $a^2 + b^2 = c^2$

$5^2 + 5^2 = c^2 \Rightarrow 25 + 25 = c^2 \Rightarrow 50 = c^2 \Rightarrow c = \sqrt{50} = 5\sqrt{2} = 5 \times 1.4 = 7$

Perimeter of triangle $= a + b + c = 5 + 5 + 7 = 17$

11) Answer: D

Range= Largest Value – Smallest Value → R $= 16 - (-4) = 20$

12) Answer: C

To start, assign the variables to unknowns, known values to constants, and relate them by the relations between the variables and constants. Then

$7w + 60 = 198$

13) Answer: B

Use percent formula: part $= \frac{percent}{100} \times$ whole

Whole $= 10 + 11 + 4 = 25$

$10 = \frac{\text{percent}}{100} \times 25 \Rightarrow 10 = \frac{\text{percent} \times 25}{100} \Rightarrow 1{,}000 = \text{percent} \times 25 \Rightarrow \text{percent} = \frac{1{,}000}{25} = 40$,

Therefore United States win 40% of medals.

14) Answer: A

19 physics teachers to 760 students is 19:760, 1:40

15) Answer: D

At least and Minimum – means greater than or equal to

At most, No more than, and Maximum – means less than or equal to

More than – means greater than

Less than – means less than

Then, P ≤ 45

16) Answer: C

If we've rounded one value up and one down, the sum we have obtained is a correct estimation.

If both summands have been rounded up and the amount of rounding is greater than 50, we subtract 100 from the estimate.

If both summands have been rounded down and the amount of rounding is greater than 50, we must add 100 to the estimate.

17) Answer: B

The formula for distance traveled over a period (time) is: $d = r \cdot t$

where d is the distance traveled, r is the rate of travel and t is the time.

$r = \frac{-42}{30} = -1.4 \rightarrow d = -1.4 \times 55 = -77$ feet

Or use proportion: $\frac{-42}{30} = \frac{x}{55} \rightarrow 30x = 55 \times (-42) \rightarrow x = \frac{55 \times (-42)}{30} = -77$ feet

18) Answer: C

The general from of scientific notation is: $a \times 10^n$ where $1 < a < 10$ and n is an integer. Then, $426{,}000{,}000 = 4.26 \times 10^8$

Common Core Math Practice Grade 7

19) Answer: D

The ratio of boys to girls in Maria Club: 18:54 =1:3

The ratio of boys to girls in Hudson Club: 1:3

1:3 same as 25: 75.

So, there are 75 girls in Hudson club

20) Answer: A

Multiply the price by the sales tax to find out how much money the sales tax will add, then Add the original price and the sales tax.

State A: $56 \times 0.0425 = 2.38$

$56 + 2.38 = 58.38$

State B: $32 \times 0.0675 = 2.16$

$32 + 2.16 = 34.16$

Then take the difference: $58.38 - 34.16 = 24.22$

21) Answer: C

$\frac{2}{9} \times 18 = 4$

$\frac{1}{4} \times 60 = 15 \, min$

$\frac{4}{15} = \frac{x}{60} \rightarrow 15x = 4 \times 60 \rightarrow x = \frac{240}{15} = 16$ Ounces

22) Answer: B

Supplementary angles are two angles that have a sum of 180°

$line \, R: 180° - 121° = 59°, Line \, S: 180° - 106° = 74°$

then in the triangle: $180° - (74° + 59°) = 47°$

$line \, P: \theta° = 180° - 47° = 133°$

23) Answer: C

$4 \times (-9) = -36$

Common Core Math Practice Grade 7

24) Answer: A

Circumference $= \pi d = 2\pi r \rightarrow 26\pi = 2\pi r \rightarrow r = \frac{26\pi}{2\pi} = 13$

Area $= \pi r^2 \rightarrow A = \pi(13)^2 = 169\pi$

25) Answer: A

$(-6) \times 14 = -84$

26) Answer: D

$45{,}364 \times 0.045 = 2{,}041.38$

27) Answer: C

$\frac{8}{33} = \frac{x}{297} \rightarrow 33x = 8 \times 297 \rightarrow x = \frac{2{,}376}{33} = 72$

28) Answer: D

$\frac{260}{5} = 52$

$\frac{520-260}{10-5} = \frac{260}{5} = 52$

29) Answer: C

Divide 0.32 by 4 to determine the number of units $= \frac{0.32}{4} = 0.08$ units

Take the units and divide by the scale. $\frac{1.28}{0.08} = \frac{128}{8} = 16$ mm

30) Answer: B

Whole numbers are a subset of integers.

All whole numbers are integers. Some, but not all, integers are whole numbers.

Integers are a subset of rational numbers.

integer can be positive, negative or zero. All integers are rational numbers.

Some, but not all, rational numbers are integers.

31) Answer: D

The diagram shows a spinner with 8 sections. The probability land on letter is

$\frac{4}{8} = \frac{1}{2}$

WWW.MathNotion.com

Common Core Math Practice Grade 7

The spinner is spun 180 times, then the prediction is: $180 \times \frac{1}{2} = 90$ times

32) Answer: A

$12\frac{3}{4} \times 10.22 = 12.75 \times 10.22 = 130.305 \; cm^2$

33) Answer: B

A linear equation is a relationship between two variables, and application of linear equations can be found in distance problems.

$d = rt$ or distance equals rate (speed) times time.

$d = 1 \times 30 = 30$, then $(1,30), (2,60), (3,90)(4,120), \ldots$

34) Answer: C

average (mean) $= \frac{sum\,of\,terms}{number\,of\,terms} = \frac{(-23)+(-27)+(-19)+3+18+(-3)+37}{7} = \frac{-14}{7} = -2°F$

35) Answer: D

$\frac{46}{0.92}$ Multiply the numerator and denominator by 100: $\frac{4,600}{92} = 50$

36) Answer: B

$m = 1 \rightarrow 5m - 2 = 5(1) - 2 = 3$

$m = 2 \rightarrow 5m - 2 = 5(2) - 2 = 8$

$m = 3 \rightarrow 5m - 2 = 5(3) - 2 = 13$

$m = 4 \rightarrow 5m - 2 = 5(4) - 2 = 18$

$m = 5 \rightarrow 5m - 2 = 5(5) - 2 = 23$

37) Answer: D

A. $3d - 20 = -35 \Rightarrow 3(-22) - 20 = -66 - 20 = -86 \Rightarrow -86 \neq -35$

B. $-25 + 2d = 18 \Rightarrow -25 + 2(-22) = -25 - 44 = -69 \Rightarrow -69 \neq 18$

C. $\frac{d}{11} + 12 = 23 \Rightarrow \frac{-22}{11} + 12 = -2 + 12 = 10 \Rightarrow 10 \neq 23$

D. $\frac{d}{2} + 35 = 24 \Rightarrow \frac{-22}{2} + 35 = -11 + 35 = 24 \Rightarrow 24 = 24$

38) Answer: A

Let's find the mode, mean (average), and median of the number of minutes for each school.

Number of Minutes for school 1: 50, 60, 70, 80, 80, 80, 90, 90, 110, 110

Mean(average) $= \frac{sum of terms}{number of terms} = \frac{50+60+70+80+80+80+90+90+110+110}{10} = \frac{820}{10} = 82$

Median is the number in the middle. Since there are an even number of items in the resulting list, the median is the average of the two middle numbers.

Median of the data is $(80 + 80) \div 2 = 80$

Mode is the number which appears most often in a set of numbers. Therefore, there is no mode in the set of numbers. Mode is: 80

Number of Minutes for school 2: 50, 50, 60, 90, 100, 100, 100, 110, 120, 120

Mean = $\frac{50+50+60+90+100+100+100+110+120+120}{10} = \frac{900}{10} = 90$

Median: $(100 + 100) \div 2 = 100$

Mode: 100

39) Answer: C

$-40 - 120d = -8(5 + 15d)$

40) Answer: C

We often describe the probability of something happening with words like impossible, unlikely, as likely as unlikely, equally likely, likely, and certain. The probability of an event occurring is represented by a ratio. A ratio is a number that is between 0 and 1 and can include 0 and 1. An event is impossible if it has a probability of 0. An event is certain if it has the probability of 1

impossible	unlikely	equally likely, equally unlikely	likely	Certain
0		$\frac{1}{2}$		1

Common Core Math Practice Grade 7

Practice Test 2

Answers and Explanations

1) Answer: C

Divided 29 by 22: $\frac{29}{22} = 1.3181818 \ldots = 1.3\overline{18}$

2) Answer: B

Day 3: $-0.84 \times \frac{3}{4} = -0.63$

Change price in days: $(89.76 + 0.92 + 0.86 + (-0.63) + (-0.86)) = 90.05$

Day 5: $91.03 - 90.05 = 0.98$

3) Answer: A

$1\,hour = 60\,min \rightarrow \frac{1}{5} \times 60 = 12\,min$

Rate: $\frac{\frac{3}{4}}{12} = \frac{x}{1} \rightarrow 12x = \frac{3}{4} \rightarrow x = \frac{3}{48} = \frac{1}{16}$

4) Answer: C

Probability $= \frac{number\ of\ desired\ outcomes}{number\ of\ total\ outcomes} = \frac{2}{7+5+10+2} = \frac{2}{24} = \frac{1}{12}$

If an event has a 0 probability this means that it can never happen

If an event has a 1 probability it will certainly happen

If an event has a 0.5 probability, it has an equal chance of happening or not happening

If an event has a probability between 0 and 0.5, then it is unlikely to happen

If an event has a probability between 0.5 and 1, then it is likely to happen,

5) Answer: D

The pattern is: $6, -3, -12, -21, -30$

6) Answer: B

$13 \div 0.5 = \frac{13 \times 10}{0.5 \times 10} = \frac{130}{5} = 26$

WWW.MathNotion.com

Common Core Math Practice Grade 7

7) Answer: C

If the price of a car is decreased by 75% then:

$25\% \text{ of } 28,000 = 0.25 \times 28,000 = 7,000$

8) Answer: D

Write the ratio and solve for x.

$\frac{1}{\frac{1}{6}} = \frac{x}{\frac{3}{4}}$ (Cross multiply) $\implies \frac{1}{6}x = \frac{3}{4} \implies x = \frac{18}{4} = \frac{9}{2}$

9) Answer: C

$90 \div 5 = 18$

$162 \div 9 = 18$

10) Answer: A

Hour: $x \to 42$ per hour: $42x$

Plus: add $(+)$, no more than: $\leq$

Then, $42x + 70 \leq 680$

11) Answer: D

$7x - 4.3y - 8x - x + 4y = 7x - 8x - x - 4.3y + 4y = -2x - 0.3y$

12) Answer: B

By grid line: $d = 4$

Or distance for two points $(0,2), (0,-2)$: $d = \sqrt{(0-0)^2 + (2-(-2))^2} = 4$

$d = 4 \to$ Circumference $= \pi d = \pi(4) = 4\pi = 12.56$

13) Answer: B

Day1: $y = 7x \to$ Day 3: $y = 7(3x)$

14) Answer: B

$x = 5 \to -3x + 4 = -3(5) + 4 = -15 + 4 = -11$

15) Answer: C

likely one-story house is: $0.36 \times 200 = 72$

likely of two-story house is: $0.45 \times 200 = 90$

16) Answer: B

Ratio of ($2\frac{2}{5}$: 144) equals to 1: 60 (both divide by $2\frac{2}{5}$)

17) Answer: D

3 hours equal 6 half hours

$6 \times 5° = 30° \rightarrow 12 - 30 = -18°F$

18) Answer: C

If the price of a printer is decreased by 46% then: $100\% - 46\% = 54\%$

$54\% \: of \: h = 0.54 \times h = 0.54h$

19) Answer: B

The median of a set of data is the value located in the middle of the data set. To find median, first list numbers in order from smallest to largest:

22, 22, 27, 29, 34, 37

Since there are an even number of items in the resulting list, the median is the average of the two middle numbers.

Median= $(27 + 29) \div 2 = 28$

20) Answer: D

$100\% - 30\% = 70\%$ eat nonorganic food

$70\% \times 160 = 0.70 \times 160 = 112$

21) Answer: A

Each hour is 60 minutes, so we have $2 \times 60 = 120$ minutes of workout time for Asher.

We subtract off the jogging and playing volleyball time to get the time Asher playing billiards: $120 - 32 - 58 = 30$ minutes

$percent = \left(\frac{part}{whole}\right) \times 100 \rightarrow percent = \left(\frac{30}{120}\right) \times 100 \rightarrow percent = 25\%$

22) Answer: C

$3\frac{5}{13} \times \frac{-5}{13} = \frac{44}{13} \times \frac{-5}{13} = -\frac{220}{169} = -1\frac{51}{169}$

Common Core Math Practice Grade 7

23) Answer: C

The reflection of the point (x, y) across the y-axis is the point $(-x, y)$.

If you reflect a point across the y-axis, the y-coordinate is the same, but the x-coordinate is changed into its opposite.

Reflection of $(-1, 4) \rightarrow (1, 4)$

24) Answer: A

Divide 3 by 8 and then times 100: $\frac{3}{8} \times 100 = 37.5\%$

25) Answer: B

$3(x + 8) + (2x - 9) + 40 = 180 \rightarrow 3x + 24 + 2x - 9 + 40 = 180$

$5x + 55 = 180 \rightarrow 5x = 125 \rightarrow x = \frac{125}{5} = 25 \Rightarrow x = 25°$

$\angle B = 3(x + 8°) = 2(25° + 8°) = 66°$

26) Answer: D

1 L = 1,000 mL

1 fl oz = 29.57 mL → 1 fl oz = 0.02957 L

48.20 L = $\frac{48.20}{0.02957} \times \frac{100,000}{100,000} = \frac{4820,000}{2,957} \approx 1,630\ fl\ oz$

27) Answer: B

$3\frac{1}{2} + 3 = 6\frac{1}{2}$ ounces

$6\frac{1}{2} \times 0.60 = 6.5 \times 0.60 = 3.9$ then the cost is $3.9

28) Answer: A

Volume of a square-based pyramid: $V = \frac{1}{3} B.h$, where V is the volume and B is the area of the base. Then, $V = \frac{1}{3} \times 11.5 \times 11.5 \times 9 = \frac{1,190.25}{3} = 396.75$

29) Answer: D

$32v + 2,150 = 16,230$ (subtract 2,150 from both sides)

$\rightarrow 32v = 14,080$ (divide both sides by 32) $\rightarrow v = \$440$ cost of each table

WWW.MathNotion.com

Common Core Math Practice Grade 7

30) Answer: B

Difference of temperature is: $|t_2 - t_1| = |15 - (-8)| = |15 + 8| = |-15 - 8|$

31) Answer: C

$\angle D = \angle E = b° \rightarrow FD = FE = 4cm$

Perimeter of triangle: $4 + 4 + 6 = 14 \, cm$

Actual triangle: $14 \times 3\frac{1}{4} = 14 \times \frac{13}{4} = \frac{182}{4} = 45\frac{1}{2} \, cm$

32) Answer: A

Use formula of rectangle prism volume.

V = (length) (width) (height) $\Rightarrow$ V = $15 \times 5.6 \times 10.75 = 903 \, ft^3$

33) Answer: B

$-6x + 10 > -20$ (Subtract 10 from both sides)

$-6x > -30$ (Divide both side by -6, remember negative change the sign)

$x < 5$

34) Answer: D

A. does not consider that the manager's weight could be different from the mean weight and would change the value of the mean

B. does not consider that the nth heaviest employee (in the middle) could be different weight and adding a weight that is greater than either of these weights to the data set would change the value of the median

C. does not consider that the manager could be the same weight as the employee and that this weight could be the new mode

D. corrects.

35) Answer: A

Supplementary angles are two angles with a sum of 180 degrees.

$\alpha + \beta = 180°$ and $\beta = 110° \Rightarrow \alpha = 180° - 110° = 70°$

complementary angles are two angles with a sum of 90 degrees.

$α + γ = 90$ and $α = 70° \Rightarrow γ = 90° - 70° = 20°$

36) Answer: B

The ratio of boy to girls is $3: 4$. Therefore, there are 4 girls out of 7 students. To find the answer, first divide the number of girls by 4, then multiply the result by 7.
$20 ÷ 4 = 5 \Rightarrow 5 × 7 = 35$

37) Answer: D

A. Expenses in Year 2: $1,600 \to 1,600 + 300 = 1,900 \neq$ Year 3

B. Income in Year 1: $1,600$ and 20% of $1,600 = 320 \to 1,600 + 320 = 1,920 \neq$ income of Year 5

C. Incomes year 2, 3, and 4: $1,800 + 1,500 + 1,800 = 5,100$

Expenses year 2, 4, and 5: $1,600 + 1,500 + 1,700 = 4,800 \neq 5,100$

D. Half of Year incomes 4: $\frac{1,800}{2} = 900 < 1,300$ Expenses in Year 3

38) Answer: C

$(3n - 9) - \frac{1}{3}(8 - 6n) + \frac{5}{3} = 3n - 9 - \frac{8}{3} + 2n + \frac{5}{3} = 5n - 10$

39) Answer: A

There are 15 total marbles $(6 + 3 + 6)$.

Probability $= \frac{number\ of\ desired\ outcomes}{number\ of\ total\ outcomes} = \frac{3}{15} = \frac{1}{5} = 0.2$

40) Answer: C

If you ever need to find the percentage of something you just times it by the fraction. So, all you need to do to figure this out is to find 35% of 13.06 which is approximately 4.57.

"End"

www.ingramcontent.com/pod-product-compliance
Lightning Source LLC
LaVergne TN
LVHW061308060426
835507LV00019B/2064